Self-Sufficiency

Home Smoking and Curing

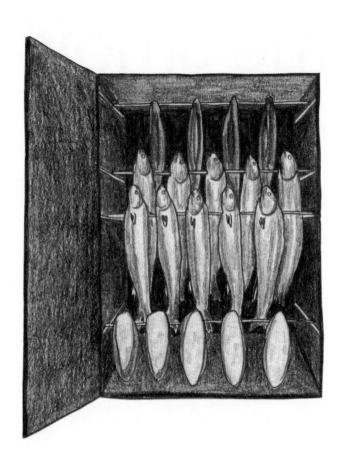

Self-Sufficiency

Home Smoking and Curing

Joanna Farrow

Skyhorse Publishing

Skyhorse Publishing books may be purchased in bulk at special discounts for sales promotion, corporate gifts, fund-raising, or educational purposes. Special editions can also be created to specifications. For details, contact the Special Sales Department, Skyhorse Publishing, 307 West 36th Street, 11th Floor, New York, NY 10018 or info@skyhorsepublishing.com.

Skyhorse® and Skyhorse Publishing® are registered trademarks of Skyhorse Publishing, Inc.®, a Delaware corporation.

Visit our website at www.skyhorsepublishing.com.

10 9 8 7 6 5 4 3 2 1

Library of Congress Cataloging-in-Publication Data is available on file.

ISBN: 978-1-61608-848-4

Printed in China by Toppan Leefung Printing Ltd

Note:
The author and publishers have made every effort to ensure that all information given in this book is safe and accurate, but they cannot accept liability for any resulting loss or damage to either property or person, whether direct or consequential or however arising.

CONTENTS

INTRODUCTION

When we experiment with any new cooking skills there is always the added excitement that comes with an element of trial and error, but the satisfaction of making your first home-cured bacon or air-dried ham cannot be underestimated—it is totally thrilling! This book provides the introduction you need to the compelling, addictive, and satisfying skills of home curing and smoking. Easy to follow techniques and recipes show how delicious cured and smoked foods can be achieved without the need for hi-tech equipment or expensive ingredients. Anyone who likes the tastiest bacon, most delicious smoked fish, or perfectly cured salami will enjoy it all the more with the knowledge that it is of good provenance, well prepared, and made by you.

The art of home curing stems from the ancient and necessary skills developed when humans had to hunt and gather their food. The basic raw materials of meat and fish, salt, fire, and air were used, as they still are today, to cure, dry, smoke, and cook. These skills remained vital to providing adequate food supplies up until the advent of refrigeration and freezing, which then took over as the main forms of long-term preservation.

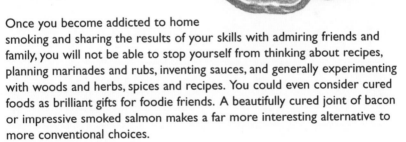

Unlike freezing, curing changes texture and flavor by infusing foods with salt, spices, sugar, herbs, and other aromatic ingredients, often enhanced by the additional flavor of hot or cold smoking. It's the ultimate "slow food" and its popularity is owed to the revolution against fast food and to our increasing interest in wanting to eat the best produce available. What will be surprising for the first-time curer is how simple the process really is. A piece of belly pork rubbed with salt and chilled for several days becomes bacon. It's as simple as that! Not only that, but it is better and cheaper bacon than most you will find in the shops.

Once you become addicted to home smoking and sharing the results of your skills with admiring friends and family, you will not be able to stop yourself from thinking about recipes, planning marinades and rubs, inventing sauces, and generally experimenting with woods and herbs, spices and recipes. You could even consider cured foods as brilliant gifts for foodie friends. A beautifully cured joint of bacon or impressive smoked salmon makes a far more interesting alternative to more conventional choices.

Curing

Curing is the preservation process in which any meats and fish are immersed in salt or brine, sometimes with additional flavorings like sugar, herbs and spices, beer, wine and cider. Salt, used either neat for a dry cure or with liquid for a brine cure, is the most essential curing ingredient and works on the basis that it draws out moisture from meat tissue, firms up texture, and in the process makes meat or fish more resilient to bacteria.

Ingredients

In theory, curing can preserve any type of meat or fish. In practice it's pork that is most widely used, partly because of its shorter keeping time as a fresh meat and also because it has a naturally high fat content, which helps to keep it succulent and well flavored. In addition, pigs are farmed or kept for food almost all over the world.

To get the best results from your home curing it is important to start with the best basic ingredients you can find. Quite simply, the better the ingredients, the better the final flavor will be.

Meat

For top-quality meat, buy from a good butcher, farmers market, farm shop, or farmer friend. This is not to say that all supermarket meat is inferior, but much of it is and it's harder to judge what is good or not. If you order via the butcher, you can feel more assured that the meat is very fresh (which is vital) and you'll be able to order decent-sized cuts of the exact weight you require.

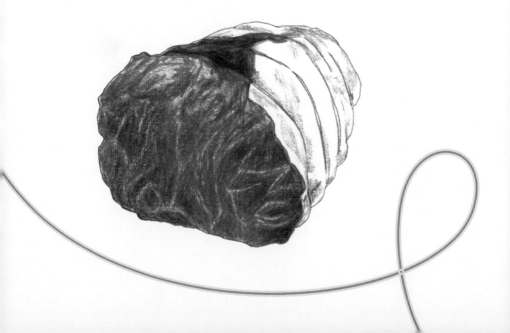

Fish

Choose fish with bright
eyes that are neither
sunken, dull, or cloudy.
The skin should have a glossy sheen
and the fish, whether whole or filleted,
should look firm and plump, rather than
limp and grey. It should also smell of the sea,
not a tired fish counter. Take care when buying oily fish such as
mackerel and sardines as they deteriorate very quickly.

Salt

Fine, pure table or kitchen salt is used for curing as it can be rubbed
thoroughly and evenly into the nooks and crannies of meat and fish.
Sea salt flakes should not be used (unless otherwise stated) as they
are too coarse.

Saltpeter (Potassium Nitrate)

Saltpeter has a toxic effect on bacteria. During the curing process it also
gives cured meats an attractive pinkish tinge and has a very slight effect
on flavor. It is still used in the curing industry, though there are now
strict controls on the amounts used because of concerns over its effects
on health. Although you might come across curing recipes that call for
saltpeter, you'll no longer be able to buy it over the counter unless you
have a special licence. It's now only available in curing salts.

Curing Salts

Curing salts can be bought through online suppliers (see page 124).
They contain salt, saltpeter, and various other flavorings and are used
in the same way that you'd use regular salt for curing, i.e., in dry curing
or making into brine solutions. If you don't want to experiment with your
own flavorings and are keen for your cured meats to have a pink tinge
(which you won't get without it), then curing salts are a good alternative
to making your own cures. Follow the supplier's directions for usage.

Spices

Historically, strong-flavored spices were added to foods being preserved to disguise the flavor of deteriorating meat. They do not actually preserve the meat, but punchy flavors like chilli, cinnamon, cumin, coriander, juniper, ginger, and cloves make lovely additions to various cures. This is an area for experiment and personal preference so you can have fun trying out different combinations. Blending spices for dry rubs, marinades, and sauces is one of the accompanying pleasures of the curing and smoking process.

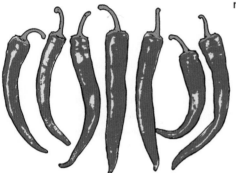

Sugar and Sweeteners

Sweet cures might include white or brown sugar in the form of muscovado or molasses, or maple syrup and honey. Sugar alone cannot be used as a curing agent, though it can be added to create a delicious flavor balance between salty and sweet.

Equipment

Don't let your initial enthusiasm propel you into a buying frenzy of specialist equipment that might not get regular use. At the most basic level of home curing, all you need is a suitable container large enough to contain your chosen cut of meat. The following list incorporates all the useful items you might want to collect over time, as well as a few basics that you may well have already. For more information on equipment, check out the list of suppliers on page 124.

- **Plastic boxes** ranging in size from food containers to larger storage boxes are ideal for all types of curing and are long lasting. It is best to get rectangular ones so pieces of meat can fit more snugly. For fridge-curing, measure up the space you have in the fridge for a suitable box before buying so you'll know how large a piece of meat to order for fitting the box. Bear in mind that you'll need a bit of space around the meat so it can freely absorb cures.

- A large, clean **wooden box** such as a wine box is useful for dry curing a whole leg of pork, or a large quantity of bacon when you might want to stack layers of belly or loin joints. For curing large joints prior to air drying you'll need to make several drainage holes in the base of the box, then raise the box on sturdy blocks over a large tray or other watertight container to catch the juices.

- **Ceramic, glazed earthenware, or glass containers** make suitable curing containers, particularly for smaller cuts of meat. Never cure meats or fish in metal containers as the cure can corrode the metal, which taints the food.

- A large **glass, plastic, or ceramic mixing bowl** is required for mixing large quantities of chopped or minced meats for making salami or chorizo.

- **Meat hooks** are useful for hanging meats to dry after curing and to air dry. Use stainless steel ones so they don't taint or spoil the meat.

- **String** that's fairly fine but strong is used for tying salami and chorizo ends and for hanging.

- **Muslin** cloth or stockinette can be used for wrapping pieces of meat for air drying. It allows the meat to breathe while protecting it from flies and other unwanted pests.

- **Kitchen scales** with a wide weight measurement range are vital for accurately measuring out ingredients, from large quantities of salt to smaller amounts of herbs and spices.

- **Weights** such as old-fashioned scale weights, unopened food cans, or rocks and stones from the garden (or garden center) are used for pressing pieces of meat or fish. If using rocks or stones, make sure you thoroughly scrub them first.

- Small **sharp knives** are essential for dicing or slicing. A larger sharp knife is adequate for the novice curer in order to thinly slice bacon or salami before going to the expense of buying a slicing machine.

- **Slicing machines** vary considerably in price and quality. While they provide the only true way of cutting wafer thin slices of meat after curing, they're a luxury rather than essential.

- **Sausage machines** make light work of stuffing salami and chorizo. Types include economical hand-held syringe or lever models through to more substantial electrical ones.

- A large **plastic** or **stainless steel funnel** can be used as a cheaper alternative to a sausage machine for filling sausage skins and ox runners. Make sure the funnel end is about ⅝ in (1.5 cm) in diameter, otherwise it will be too difficult to push the meat through.

- **Brine pumps**, used to accelerate brining times of wet cured meats, can be as basic as a small syringe that forces the brine into the center of the meat. Pricier pumps have a long needle with holes along its length allowing even penetration into the meat.

Dry curing

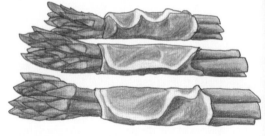

In its simplest form, dry curing is the rubbing of salt into a piece of pork such as belly or loin to make a simple bacon, or into a fillet of fish to dry it for preserving prior to cooking. A lengthier form of dry curing is to salt a joint under a heavy weight to preserve and extract a certain amount of water, then air dry it for several months so further moisture is extracted right through to the center of the meat. This produces a ham that does not need to be cooked. Famous examples of these hams are the delicious Italian Prosciutto and Spanish Serrano hams. Hugely expensive to buy, a homemade version is well worth attempting and can produce good results provided you're well prepared—and patient! The only requirements are time and a good quality joint of meat to make your curing exploits worthwhile.

Curing times

The longer the meat is cured, the saltier it will become and the more prolonged its shelf life will be. The bacon recipe on page 22 is cured for five days and will store in the fridge for two weeks. A piece of pork that's cured for two weeks will keep for up to a month in the fridge but may well taste overly salty. You can cut off pieces as required and reduce the saltiness by soaking it overnight in water, changing the water several times as you would a bought bacon joint, before using. Home-cured bacon can also be frozen for several months.

Dry curing without refrigeration

During the summer, the only fail-safe way to cure meats is to use the fridge. It is still possible to achieve good results without refrigeration but the risks of the meat rotting are greater. A dry, cool larder, cellar, or basement can also be used successfully. If these aren't available and you don't have any fridge space, try using a large, cool box, placing freezer blocks under the container of meat and replacing them daily. During the winter, garages or garden sheds can make useful sites to store your curing meats but always use containers with lids to protect the meat from unwelcome visitors.

Pork cuts

Almost any part of the pig can be made into dry-cured bacon but the best cuts to use, both in terms of easy curing and slicing, are the belly and loin. Ask the butcher to remove the skin and bones for you, or if you prefer to dry cure bacon with the skin attached as a more traditional method, follow the basic recipe allowing an extra 24 hours curing. Belly pork is a good cut to start with as it's inexpensive and is a thin cut, therefore curing more quickly.

It's also easier to slice using a sharp cook's knife so you needn't buy expensive slicing machines. Once cured, pork loin is known as back bacon whereas belly pork is called streaky.

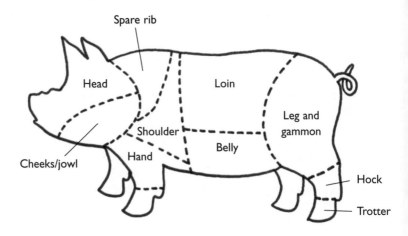

Dry-cured ham
It's possible to dry cure a whole or half leg of pork to make a dry-cured ham for boiling. Use a tunnel-boned leg or half leg of pork and pack it in salt for three to four weeks, see page 59 for method. After this time you can boil it as you would a brined leg of pork.

Curing more than one piece at a time
If you want to cure several pieces of pork at the same time, use the quantities and method as in the basic recipe on page 59 for each piece and stack them in the container, making sure that each piece is thoroughly salted before stacking. Rearrange the pieces daily.

Adding flavors

Sugar, herbs, and spices can be added to the salt when dry curing to enhance the flavor. Use approximately 20% sugar to salt i.e., 1 oz sugar to every 5 oz salt, and stir in additional flavors such as dried crushed chilli, juniper berries, fennel seeds, ground black pepper, chopped rosemary, thyme, sage, or crushed bay leaves. Seek inspiration from artisan bacon flavorings at farmers markets and supermarkets, but keep the flavorings quite simple, bearing in mind that the dominant flavor will be of the bacon, not the flavorings added to the cure.

Using dry-cured bacon

Your home-cured bacon can be cooked and used in exactly the same ways you would bought bacon. When frying you'll notice that it won't emit a lot of water as some bought bacon does and the pieces will easily crisp and color. It will also be darker and less pink due to the absence of saltpeter. Slice, dice, or roughly chop depending on your chosen recipe (see page 37).

How to make low-salt bacon

Low-salt bacon is just as easy as making regular bacon and is made overnight, however it'll only keep for a couple of days in the fridge.

Use ¼ in (5 mm) thick slices of skinned belly or loin of pork, either cut by the butcher or with a large sharp knife. For 1 lb 2 oz (1 kg) pork use two teaspoons fine salt, mixed with one teaspoon caster sugar. Add any other flavorings you like, such as freshly ground black pepper, a sprinkling of finely chopped sage, thyme, or rosemary, or crushed coriander seeds. Sprinkle the rub very lightly over the pork slices and place in a shallow non-metallic dish. Cover loosely with plastic wrap and chill overnight. Rinse off the salt, pat dry on paper towels, and use as you like!

Dry-cured streaky bacon

Making bacon at home is the easiest of all the home curing methods and will produce good, foolproof results. This recipe is the simplest of all the cured meats, it's accessible to everyone, and requires no special equipment or ingredients, just regular salt and a decent piece of pork. Use it as an introduction to home curing. You'll be so pleased with the results and will no doubt want to experiment further. The amount made is reasonably small but you can easily make larger quantities, it's a question of how much bacon you get through. For additional flavor the bacon can be smoked (see page 92).

Ingredients

2 lb 2 oz (1 kg) belly pork, skinned

Approx 2 oz (50 g) salt

1 Thoroughly dry the pork on all sides with paper towels to remove any moisture that has accumulated on the surface. Using your fingers, rub half the salt into the surface of the meat on all sides, making sure you get into any cracks and crevices.

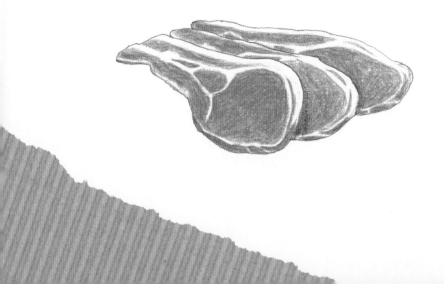

2 Once thoroughly salted, place the pork in a clean plastic or wooden container (see Equipment, pages 13–15, for types of containers) and place in a cool place, or refrigerate overnight.

3 The next day you'll find that a small pool of liquid has accumulated in the base of the container. Drain this off, pat the meat dry, and rub with the remaining salt. Return to the container and store in the fridge or cool place for a further 4 days, turning the pork daily so that it cures evenly.

4 Rinse the bacon in several changes of water to remove all excess traces of salt. Dry thoroughly on paper towels. The bacon is now ready to use but should, ideally, be hung for a further 4–5 days to dry out before cooking or smoking.

5 Make a hole through the pork, about ¾ in (2 cm) away from one corner, by pushing the tip of a small, thoroughly clean knife through the meat and twisting the knife slightly until you have a hole that's large enough to thread a piece of string through. Cut a length of string and push it through the hole. Use the string to hang the meat in the fridge. It might drip a tiny amount of moisture so shouldn't be stored over other foods, just in case. If necessary, place a small tray or plate under the bacon. Store for up to two weeks or, for longer term storage, cut into two or three pieces and freeze so you can thaw pieces overnight as required.

Sweet-cured bacon with rosemary, mustard and molasses

Molasses sugar gives this bacon a dark tinge and a full, hearty flavor.

1 Put the mustard seeds in a small, dry frying pan and heat gently until the seeds start to pop. Grind using a coffee grinder reserved for spice grinding or by using a pestle and mortar. Pull the rosemary leaves from the stalks and finely chop.

Ingredients

2 tbsp mustard seeds

⅓ oz (10 g) rosemary sprigs

3½ oz (100 g) salt

¾ oz (20 g) molasses sugar

2 lb 2oz (1 kg) piece loin of pork, skinned

2 Mix together the mustard seeds, salt, sugar, and rosemary.

3 Rub half the mixture all over the surface of the meat and complete curing following the recipe for dry-cured streaky bacon (pages 20–21).

4 Once cured and hung for several days, the bacon is ready to eat. It can also be cold-smoked to add additional flavor following the method on page 92.

Slicing bacon

All you need to slice the bacon is a good, sharp cook's knife and chopping board. Place the bacon on the board fat side face down and cut slices as thinly as possible. These slices won't be as regular or thin as bought sliced bacon but the flavor will be more delicious and their irregularity is part of their homemade charm! If you intend to make more bacon, you might want to invest in a meat slicer (see page 15). These are generally fairly expensive and only worth buying once you're convinced you want to do plenty of home curing.

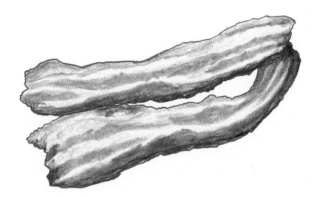

Flavor variation

For a lighter, mildly spiced, sweet cure, try maple sugar with coriander and lemon. Finely crush 3 tbsps coriander seeds and mix with 3½ oz (100 g) salt, the finely grated zest of 2 lemons, and ¾ oz (20 g) maple sugar. Use as above.

Cured fish

There is as much flexibility in the techniques used for curing fish as there is for meat and the basic principles remain the same (i.e., preserving by drawing out the moisture using salt). Any fish can be cured but those that work best are larger fish with plump flesh and large flakes that retain a good texture. Smoked salmon, which is usually brine cured, is the most popular and most available cured fish, but there are many other delicious alternatives such as salted and pickled fish.

Salt fish

White fish, perhaps most famously cod, has been salted and dried for centuries as a form of long-term preservation. Visits to fish markets throughout the world reveal a feast of fish, whole or filleted, that have been salted and dried for long-term storage. Nowadays we salt fish mainly to intensify its flavor, firm its texture, and make all the delicious dishes that

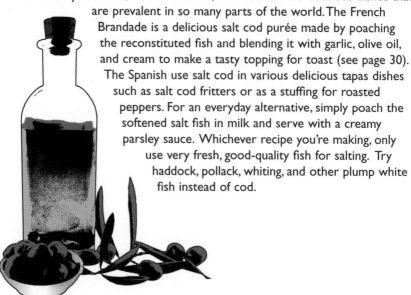

are prevalent in so many parts of the world. The French Brandade is a delicious salt cod purée made by poaching the reconstituted fish and blending it with garlic, olive oil, and cream to make a tasty topping for toast (see page 30). The Spanish use salt cod in various delicious tapas dishes such as salt cod fritters or as a stuffing for roasted peppers. For an everyday alternative, simply poach the softened salt fish in milk and serve with a creamy parsley sauce. Whichever recipe you're making, only use very fresh, good-quality fish for salting. Try haddock, pollack, whiting, and other plump white fish instead of cod.

Storing salt fish

The longer the fillets are submerged in the salt, the saltier the flavor will be. After three days the fish can be drained and used, but you can keep it submerged for up to two weeks. After this time the fish can be drained and dried for longer-term storage.

Drying salted fish

Remove the fish from the dissolved salt but do not rinse it. Pat dry on paper towels and make a small hole about ½ in (1 cm) away from the end of the fillets with the tip of a thoroughly clean small sharp knife. Thread with string and hang the fillets in the fridge or a very cool place such as a cellar, larder, or outbuilding. The fish will become hard and turn to a matte, pale, cream color, which will keep for several months if conditions are right. Check frequently to see how it is drying. To reconstitute, soak the fish in cold water for 24–48 hours, changing the water once or twice until softened.

Salt fish fillets

Salting fish is practised all over the world to intensify flavor and preserve. Don't forget to use absolutely fresh fillets.

Ingredients

2 large fillets of pollack, haddock, or cod, skin attached

Plenty of fine salt

1 Lay the fillets out on a board and check over for any small bones, removing them with tweezers if necessary. Pat dry with paper towels.

2 Sprinkle a ½ in (1 cm) layer of salt into a ceramic, plastic, or glass container large enough to hold the fillets without having to bend them. (If necessary trim the fillets to fit the container). Lay one fillet on top, skin side up. Sprinkle with another thick layer of salt and position the second fillet, skin side up.

3 Sprinkle with another thick layer of salt. Place a small plate or lid that's small enough to fit inside the container and weight the fish with kitchen scale weights or food cans. Chill for at least three days, uncovering the fish daily to check that it's still submerged in the salt solution. If necessary, sprinkle in more salt.

4 After three days, drain off the salt, which will have dissolved completely to a liquid, and pat the fillets dry on paper towels. If cooking the fish immediately or within 24 hours, test for saltiness by cooking a small piece in a little milk or water. If too salty, soak it for one to two hours in cold water first. Alternatively, dry the fish for longer term storage.

Pickled fish with Asian spices

The idea of pickling fish is a strange notion to some of us but the results are absolutely delicious. Most fish, both white and oily, work well when bathed in a sweet pickling vinegar with spices and other seasonings. Chill for at least 24 hours before serving so the fish has time to "cook" in the marinade. It's lovely served with warmed grainy bread or flatbreads.

Ingredients

Fillets from 8 herrings or small mackerel

2 red onions, very thinly sliced

2 tsp coriander seeds, crushed

2 tsp cumin seeds, crushed

2½ tsp sea salt

3½ oz (100 g) caster sugar

½ tsp dried crushed chillies

3 cloves garlic, thinly sliced

7 fl oz (300 ml) white wine vinegar

2 tbsp chopped coriander

2 tbsp chopped mint

1 Check over the fish for any small bones, removing with tweezers if necessary. Place the fillets in a large shallow dish and scatter with the onions.

2 Heat the coriander and cumin seeds in a small, dry saucepan for 30 seconds. Add the salt, sugar, chillies, garlic, and vinegar. Bring almost to a boil and remove from the heat.

3 Leave to stand for five minutes then stir in the coriander and mint and pour over the fish. Cover loosely with plastic wrap and chill for up to a week. Serve at room temperature with bread or flatbreads for mopping up the juices.

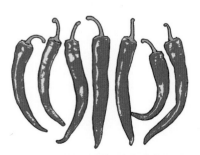

Gravad Lax

This Swedish recipe is a delicious way of curing fish, short term. It's usually done with salmon fillets though it works equally well with large, chunky fillets of trout. Served in wafer thin slices, Gravad Lax has a smoked salmon-like quality that makes a delicious starter or snack and your homemade version will cost a fraction of any you might buy.

Ingredients

2 salmon fillets, each weighing about 1 lb 2 oz (500 g)

1 oz (25 g) finely chopped dill

1½ oz (40 g) sea salt

2 oz (50 g) caster sugar

3 tbsp black or white peppercorns, finely crushed

FOR THE SAUCE

2 tbsp French brown mustard

1½ tbsp caster sugar

5 tbsp chopped dill

3½ fl oz (100 ml) mayonnaise

1 Lay the fillets out on a board and check over for any small bones, removing them with tweezers if necessary. Pat the fillets dry with paper towels.

2 Lay one fillet skin side down in a shallow ceramic, plastic, or glass container in which it fits quite snugly without having to bend. Mix together the dill, salt, sugar, and peppercorns, and spoon down the length of the fish. Cover with the second piece of fish, skin side face up.

3 Cover the dish loosely with foil and place a small plate or lid that is small enough to fit inside the dish. Weight the fish with kitchen scale weights or food cans. Chill for two to three days, turning the fish twice a day so that each fillet receives the same amount of salting. The salt mixture will gradually turn to liquid during curing.

4 For the sauce, beat together the mustard, sugar, dill, and mayonnaise and place in a small dish.

5 Remove the salmon from the dish and separate the two fillets. Using a sharp knife, cut off thin slanting slices from the salmon so each slice is topped with a little dill and pepper. Serve with the sauce.

Brandade

Serve this deliciously creamy purée as a summery snack or simple starter with warmed bread. It is perfect scattered with black olives and parsley or topped with poached quail or chicken eggs.

Ingredients

1 lb 9 oz (700 g) salted white fish fillets, see page 26, soaked for 24–48 hours in cold water to soften

3 bay leaves

1 onion, quartered

½ pint (300 ml) half-and-half

3½ fl oz (100 ml) olive oil

3½ fl oz (100 ml) milk

4 garlic cloves, crushed

Squeeze of lemon juice, plus lemon quarters to serve

Plenty of freshly ground black pepper

1 Drain the fish and put in a saucepan. Cover with fresh water, add the bay leaves and onion, and bring to a boil. Reduce the heat and simmer gently for about 15 minutes or until the fish is very tender. Drain to a plate.

2 When cool enough to handle, check over the fish for any bones and pull away the skin. Put the half-and-half, oil, and milk in a saucepan with the garlic and bring to a gentle bubble.

3 Put the fish in a food processor and blend to a smooth paste. Add a little of the cream mixture and blend again until smooth. Continue blending and adding the liquid until the mixture is smooth, light, and creamy. Add a squeeze of lemon juice and black pepper to taste.

4 Turn into a bowl and serve warm (or reheat gently in a saucepan), with lemon quarters, warm crusty bread, and some olives.

Brine curing

This is the form of curing usually reserved for bacons and hams that you want to boil or bake after curing. Brine curing uses salt and similar flavorings to dry curing, except that the ingredients are mixed with liquid to create a solution for steeping the meat. It is usually reserved for pork to produce a wet-cured bacon (once generally known as green bacon), or ham and is more commonly used in cooler climates where air-drying is traditionally less successful. Salt beef is another well-known example of this technique and your homemade salt beef will be vastly superior to anything you'll buy. Brine-cured meat can be taken one stage further in the flavoring process by cold smoking, see page 90, before cooking. All brine-cured meats are cooked before eating.

Most bacon and ham we buy is brine- rather than dry-cured because the process is faster and is therefore more profitable for the manufacturers. An injector forces brine into the center of joints of meat where the steeping brine cannot reach. This speeds up the curing process and makes the meat safer to eat. Brine injectors are not essential for home curing but make a worthwhile investment if you are keen to cure large joints, for example a chunky joint over about 4 lb 6 oz (2 kg), as opposed to thinner cuts such as belly and loin.

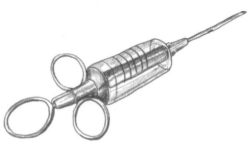

Ham or bacon?

Almost any cut of pork can be brine or dry cured to make bacon. When the meat is taken from the hind leg of the pig and then cured and cooked it becomes ham, though uncooked cured pork from the leg is also often labelled ham. Gammon is cut from the top of the leg.

Using a brine injector

Before immersing the meat in the brine, it can be injected with some of the brine solution to speed up curing. The amount injected needn't be accurately measured but as a guide you want to inject about 2–3 tablespoons of brine to every 1 lb 2 oz (500 g) of meat (so don't forget to weigh the meat first!). Fill the injector with brine and push it firmly into the meat as near to the bone (if present) as you can. Gently start pulling the injector out as you squeeze out the brine. Repeat all over the joint.

The egg test

This is the simplest method of ensuring that your brine solution contains enough salt for curing and is worth doing every time you make up a brining solution. Once you've mixed the ingredients and the salt has dissolved, place a whole uncooked egg in the bowl. The solution is salty enough if the egg bobs around near the surface. If it sinks to the bottom, or hovers somewhere in the middle, stir in a little more salt and do the egg test again.

Flavor variations

Develop your own recipes by using your own curing blends. Bay leaves, sage, rosemary, thyme, and parsley can be added to the brine but use generous amounts, roughly chopped, so their flavors have a subtle impact. Other ingredients such as garlic cloves, thinly sliced, a couple of tablespoons of crushed juniper berries, crushed peppercorns or mustard powder, or plenty of whole cloves can be added to the brine. Cider, wine, or beer can be used to replace or partially replace the water. Keep flavorings simple and avoid mixing too many flavors in one session.

Ready made curing salts, both basic and in many flavors can be used for brine curing bacon and ham, see page 124 for suppliers and follow their instructions for making up and using the salts.

Brine-cured ham

Brine cures that are used for bacon are equally suitable for curing a half or a whole leg of pork. You'll need to double up or even triple the brine quantities given, depending on the size of the joint you're using. It is worth brine injecting the meat first to aid curing. Place the pork in a container as above (a large plastic one is ideal), add the brine, and weight. Drain and rinse before cooking.

Curing times

There are no specific curing times for bacon or ham and much depends on the joint used and whether you've used a brine injector. Of course, the longer the meat is left in the brine the saltier its flavor will become. As a guide, a joint weighing 4 lb 4 oz (2 kg) will need 48 hours. Add 12 hours for each extra 2lb 2 oz of meat. If the joint has been brine injected you can halve the brining time.

Cooking brine-cured bacon and ham

Once your bacon has been brined, rinsed, and dried it is very much like a piece of store-bought bacon (only better!) and can be cooked in the same way. Before you cook the whole joint, it's worth shaving off a small piece, cooking it, and tasting for saltiness. If too salty, place in a bowl of cold water and leave to stand overnight. (You might want to do the taste test as soon as you've drained the rinsed the meat so that you're not left waiting).

Here are some ways you might use your home-cured bacon...

- Dice into lardons and fry in a pan to color before adding to tarts and quiches, pasta dishes, soups, omelettes, or salads.

- Slice as thinly as possible into bacon strips for a big breakfast fry up.

- Cut into chunkier pieces and lightly fry before adding to poultry or game casseroles, beef stews, or creamy chicken pies.

- Slice thinly, grill, then pair with white bread for the perfect bacon sandwich.

...and some ways for your home-cured ham:

• Gently poach in a pan of water with plenty of onions, carrots, celery, thyme, bay leaves, and peppercorns or star anise, allowing 20 minutes per 1 lb 2 oz (500 g). Leave in the water for 20 minutes then serve thinly sliced with a creamy parsley sauce.

• Cook as above, leave to cool in the liquid, and slice thinly for delicious sandwiches.

• Cook as above, drain the meat, and use the flavored stock to make a creamy pease porridge.

• Cook as above, remove the skin, and transfer the joint to a roasting tin, fat side up. Pack a thin layer of brown sugar over the fat or spread with a honey mustard or marmalade glaze. Bake at 425°F/220°C until the glaze is deep golden, basting with the syrupy juices regularly throughout the cooking time.

Brine-cured bacon

Use a joint of pork collar, belly, or loin for this recipe, with any bones removed. Both the brine and pork should be chilled to refrigerator temperature before immersing the pork in the brine. This recipe is intended for curing in the fridge. If you don't have room, put the container in a cold cellar or larder (or in the garage or shed during winter months), covered with a tight-fitting lid.

Ingredients

4 lb 4 oz (2 kg) piece pork

9 oz (250 g) fine salt

2 tsp ground allspice (optional)

3½ oz (100 g) caster sugar

6 cups (1.75 litres) cold water

1 Put the salt, allspice (if using), sugar, and water in a large container (one you know that the meat will fit into) and stir frequently until the salt has completely dissolved.

2 Put the container of brine in the fridge and chill to fridge temperature before lowering in the meat, also at fridge temperature.

3 To ensure that the meat stays immersed in the liquid, rest a slightly smaller container on top of the meat and weight this down with a couple of food cans or kitchen weights, making sure the meat is completely immersed. Transfer to the fridge and brine
(see Curing Times page 35).

4 Lift the meat from the brine and rinse under the cold tap to remove the excess brine. Dry the joint for a further 24 hours by placing it on a wire rack resting over a plate in the fridge. The bacon is now ready to cook, or it can be stored in the fridge for a further 7–10 days. Alternatively, cold smoke the joint before cooking.

Beer, muscovado, and fennel cured bacon

This cure has a stronger, beery flavor and is a little bit special, definitely a bacon for serving unadulterated, maybe in a doorstop sandwich with a fried egg.

Ingredients

4 tbsp fennel seeds

4 lb 4 oz (2 kg) piece pork, skinned

7 oz (200 g) fine salt

3 oz (75 g) dark muscovado sugar

5½ cups (1.5 litres) bitter ale

1 Lightly crush the fennel seeds using a pestle and mortar. Tip into a small, dry saucepan and heat gently for about 1 minute until beginning to give off their scent.

2 Tip into a large container (one you know that the meat will fit into) and add the salt, sugar, and beer. Stir frequently until the salt has completely dissolved, then leave to stand until the beer froth has settled.

3 Cure and complete as in steps 2, 3, and 4 of the basic recipe, opposite.

Brine cure for fish

You don't have to cure fish before smoking. In fact, a pile of fresh mussels or clams, or small fresh fish, particularly oily ones like sardines and sprats, are delicious cooked quickly in a hot smoker with nothing more than a scattering of herbs or oil baste. For more "serious" smoking the fish is usually brine cured first. Plan your curing time prior to smoking, bearing in mind that the fish will need to dry for several hours before being put on the rack.

To make up the brine, dissolve I lb 4 oz (550 g) fine salt in 8 cups (2.25 litres) cold water until the solution is clear. To check that the solution is salty enough, place a raw egg in the water; it should float. If necessary stir in a little more salt. Pour the solution into a suitable container. A rectangular plastic storage container is ideal though you can use glass or glazed earthenware. (Do not use metal containers.) To this basic brine you can add other ingredients such as sugar, bay leaves, fennel, rosemary, thyme, or crushed juniper berries. White wine, cider, and beer can also be used but reduce the amount of water to accommodate the extra liquid. Lower the fish into the brine and leave to brine using the guide opposite.

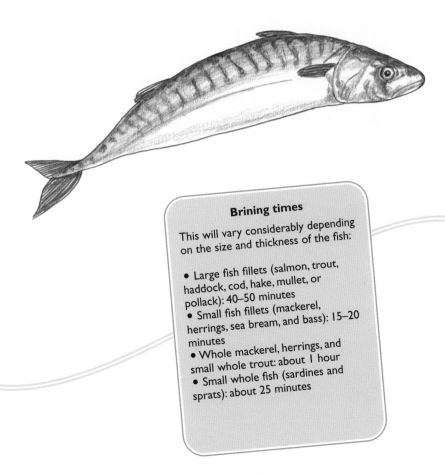

Brining times

This will vary considerably depending on the size and thickness of the fish:

- Large fish fillets (salmon, trout, haddock, cod, hake, mullet, or pollack): 40–50 minutes
- Small fish fillets (mackerel, herrings, sea bream, and bass): 15–20 minutes
- Whole mackerel, herrings, and small whole trout: about 1 hour
- Small whole fish (sardines and sprats): about 25 minutes

Wet- or dry-brined meat and fish must be dried before smoking. First pat dry lightly on paper towels, then dry further by placing the fish or meat on a wire cooling rack and leaving for several hours. This can be done at room temperature or in the fridge.

Salting beef

Steeping cheaper cuts of beef in brine along with aromatic spices preserves the meat, adds flavor interest, and tenderizes its texture when cooked. Home-cured salt beef has a delicious flavor and cooks to a meltingly tender texture. Like most brined meats, it must be soaked in water after curing to remove excess salt and requires a little planning ahead so that it's ready to cook, post brining, on the day you want to eat it.

Salt beef

You will need to start soaking the beef eight days ahead of serving.

Ingredients

9 oz (250 g) light muscovado sugar

1 lb 2 oz (500 g) fine salt

4 garlic cloves, sliced

1 teaspoon crushed mace

2 teaspoons juniper berries, crushed

2 teaspoons coriander seeds, crushed

2 oz (50 g) fresh root ginger, chopped

1 teaspoon whole cloves

4 lb (1.8 kg) piece beef brisket

1 Put the sugar, salt, garlic, mace, juniper, coriander, ginger, and cloves in a large container (one you know that the meat will fit into fairly snugly) and add 9 cups (2.5 litres) cold water. Leave to stand, stirring the mixture frequently until the salt and sugar have completely dissolved.

2 Immerse the beef in the brine, making sure there's enough brine to completely cover it. To ensure that the meat stays immersed in the liquid, rest a slightly smaller container on top of the meat and weight this down with a couple of food cans or kitchen weights. Transfer to the fridge and leave for one week.

3 Lift the meat from the brine and soak in cold water for a further 24 hours, changing the water after several hours to remove excess salt. The beef is now ready to cook, see below.

Cooking salt beef
Put the beef into a large casserole dish in which it fits quite snugly. Surround the meat with plenty of chopped carrots, onions, and celery. Add enough water to just cover the meat. Cover and cook in a preheated oven, 300°F/150°C for about three hours or until the meat is meltingly tender.

Salt beef is delicious served with horseradish or mustard mash, seasonal vegetables, and with some of the cooking juices spooned over. Leftovers are perfect for serving cold in sandwiches or finely chopping, blending with herbs and enough soft butter to bind everything together, before packing into jars for a delicious potted meat.

Pastrami

Pastrami is salt beef that has been smoked and cooked. There are various stages to making pastrami but the process is so simple to do and guarantees success. Like hot salt beef, good pastrami is so tender that it falls easily into shreds after cooking. It can also be chilled and pressed with weights for cutting into thin slices and serving with rye bread and gherkins, or whichever condiments you wish.

Ingredients

4 lb (1.8 kg) salt beef, see page 42, soaked in water for 24 hours

4 tablespoons black peppercorns

4 tablespoons coriander seeds

1 Drain the beef from the soaking liquid and pat dry on several paper towels. Slice the meat lengthways in half to make two thin joints.

2 Finely crush the peppercorns and coriander seeds. Take care not to over-grind them to a powder.

3 Place the meat on a board and rub the peppercorns and coriander into all sides of the meat until completely coated.

4 Hot smoke the pieces of meat, either using a kettle barbecue or hot smoker for 1½–2 hours until the surface is dark brown. If using a barbecue, cook the meat to one side of the heat source, turning it frequently so it doesn't start to crisp.

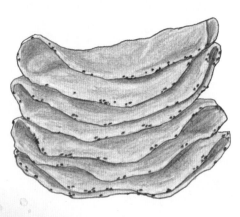

5 Preheat the oven to 250°F/120°C. Transfer the meat to a wire rack over a roasting tin and pour in a 1¼ in (3 cm) depth of boiling water. Cover the tin with a loose tent of foil, making sure the foil is secured tightly around the tin to trap the steam.

6 Cook in the preheated oven for 3½–4 hours or until completely tender. Serve hot and cut into chunky slices or remove from the heat and leave to cool.

Pressing pastrami for slicing

To thinly slice cold pastrami for sandwiches it's easiest and traditional to have weighted it down first in order to compress the meat. Place the pieces of meat in a shallow dish, cover loosely with foil and rest a small container on top. Weight with a couple of cans or kitchen weights and chill overnight before slicing.

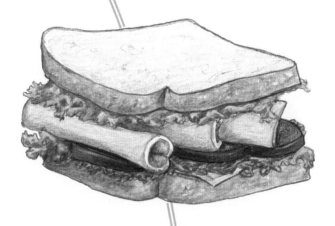

Curing tips

• Have everything ready for curing so that once you get the meat or fish home it's not sitting around in the fridge for days while you search out ingredients and equipment.

• Once the cured bacon has been rinsed and dried, shave off a small piece and fry to check its saltiness. If too salty you can soak the bacon in cold water for six to eight hours to make it more palatable.

• A white powdery mold occasionally forms on bacon during storage. This is not harmful and can be easily wiped off with a vinegar-soaked cloth.

• During brine curing give the liquid an occasional stir in case the flavorings have settled.

• Only use good quality meat or fish for home curing. You'll probably need to order this in advance if making a large quantity.

• Every time you experiment with curing, takes notes and observe how the process is going. You may well have different results depending on the time of year so keep a diary of your successes. Also record any flavorings you try for the first time. It's easy to remember what you put in a salami once you've just filled the casings, not so easy three months later when you're tucking in!

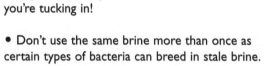

• Don't use the same brine more than once as certain types of bacteria can breed in stale brine.

• Curing meats should have a pleasant smell, like you might get walking into a delicatessen. Keep an eye on your drying meats and fish; if they start to turn black or smell bad, discard them. Sadly you'll have to start again.

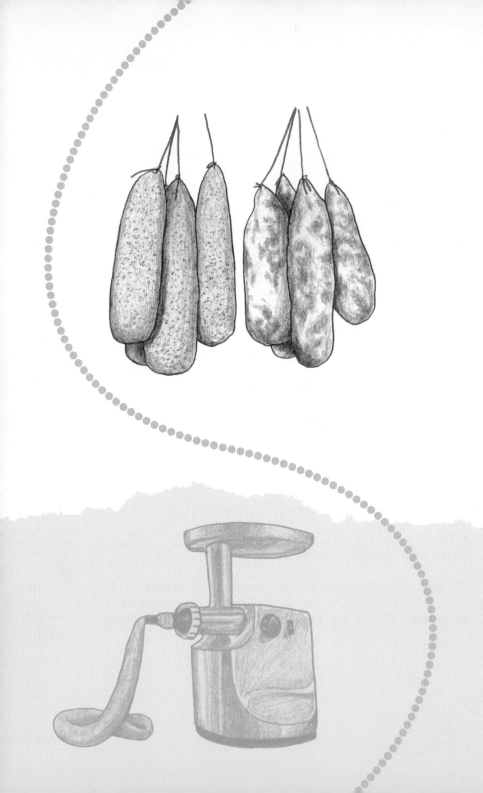

Air drying

Air-dried meats have a totally unique texture and flavor due to the lengthy process of hanging to develop flavors. Prosciutto and salami are classic examples of air-dried meats, preserved through salting and hanging in a dry, well-ventilated place for several weeks (and up to eight months for large joints) until the meat has lost all of its moisture and has developed its delicious flavor.

Air-dried meats

All bought air-dried meats are expensive, particularly those cut from a whole air-dried joint, so it's well worth trying to make your own. Success depends on accurately following the right balance of ingredients (i.e., the ratio of salt to meat), keeping your ingredients and equipment in a thorough state of cleanliness and, more importantly, on the drying itself. The challenge in some climates is finding a sufficiently well-ventilated site where dry air freely circulates. Many recipes suggest an outbuilding where doors or windows can be left open, a veranda, well-ventilated cellar, or even carport. If none of these is available, drying can work successfully if the meat is left hanging by an open window where there is a cool draft.

Salami and chorizo

Homemade salami and chorizo can be absolutely delicious. You can experiment with your own flavor additions and choose how chunky or fine you want the meat to be. The correct percentage of salt to meat is vital. Use 1 oz (25 g) fine salt to 2 lb 4 oz (1 kg) of meat. At least ten percent of the meat should be fat. Ask the butcher for "back fat," which is pure fat from under the skin of pork. The rest can be lean pork from any cut. After drying, salami and chorizo will keep well in the fridge for a couple of weeks. Any that you want to keep for longer can be frozen.

Casings

Both natural and synthetic casings are available. As with sausage making, it's best to use natural if you have the choice. You can use sausage skins, which will produce a thinner salami, or beef casings, sometimes called "ox runners," which can be filled to make a thicker salami. A thinner salami will dry quicker so is a good choice for a first attempt. To prepare the casings simply soak them in cold water for 10–15 minutes until malleable and soft. Drain from the water and use while still wet but not dripping. Any leftover casings can be frozen for another time.

Filling casings using a sausage maker
This is the easiest way to fill the casings. Simply thread a length of casing onto the thickest nozzle attachment of the machine, leaving the end of the casing untied. Load the sausage machine with the meat mixture and slowly squeeze it though with one hand to fill the casing, supporting the casings as they fill with the other hand so it doesn't buckle up. Once filled to the required length, remove from the nozzle, leaving plenty of the casing ends left for tying. Tie knots as close to the filling as you can, then tie with string for hanging and drying.

Filling casings by hand
Filling cases by hand is slightly trickier than using a sausage machine but is more cost effective for first-time curing. Prepare the casings as above. Use a large funnel that has a spout around ⅝ in (1.5 cm) in diameter and thread the casings over the spout. Fill the funnel with the sausage mixture and press it through into the casings.

Hanging

Once you have found a suitable place to hang meat, you'll need to devise a method of hanging it. A sturdy curtain rail is ideal if using a window; otherwise, you will need to secure hooks on a small, sturdy rail. Place hooks over the rail for looping string-tied sausages over or use the string to secure the meat. Hang meat or sausages, making sure they are not touching each other or the wall.

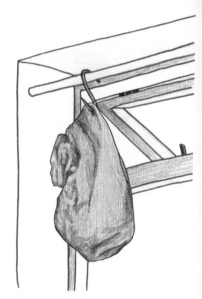

Drying times for salami

Drying times will vary but can be as little as two to three weeks. Check for readiness by squeezing the meat in the center. When it is firm and barely gives at all, take one salami and slice off the end. If still soft in the center, you will need to hang them for longer. Bear in mind that the longer you hang the meat, the dryer it will become. It's quite easy to "over dry" a salami, making the meat slightly too chewy to enjoy.

Temperature

The temperature can of course be very difficult to control. Ideally it should not exceed 53°F/12°C, so keep a regular check on it. If you are concerned about a sudden heatwave (particularly if going away), put the meat in the fridge until the temperature cools down again.

Tips

• Try adding other flavorings such as mustard, celery, caraway, cumin or coriander seeds, hot paprika, chopped pistachio nuts, and extra garlic.

• If experimenting with your own flavorings, take a small piece of the mixture and fry until cooked to test the flavor. The texture and flavor won't be exactly the same as your finished sausage but it certainly gives you an indication.

• Sausage casing is edible but can also be peeled away before slicing.

Salami

You can make one large salami but for a first attempt it's more practical, quicker, and more likely to succeed if you make small ones. This recipe makes three salami, each weighing about 6 oz (175 g). The drying time will depend on the ambient temperature and will take from two to four weeks.

1 Finely chop or mince the pork and fat. This is a matter of personal choice, if you like a coarse-textured salami, finely chopping the meat is sufficient. For a finer texture use a food processor or mincer to grind the meat to a consistency that is similar to bought minced meat.

Ingredients

2 lb (900 g) lean pork, diced

3½ oz (100 g) pork fat, diced

1 oz (25 g) fine salt

1 tsp fennel seeds, lightly crushed

1 tsp ground paprika

1 garlic clove, crushed

Plenty of freshly ground black pepper

Sausage skins or beef casings (see page 51)

2 Put the meat in a large bowl and add the salt, fennel seeds, paprika, garlic, and pepper. Use your (scrupulously clean) hands to thoroughly work the ingredients together so the flavors are evenly combined.

3 Prepare and fill the casings (see page 51).

Chorizo

Air-dried chorizo can be sliced and eaten raw like salami or added to meat and shellfish dishes, to which it adds a delicious garlicky, spicy flavor and vibrant color from the generous addition of paprika. Smoked paprika, now widely available in supermarkets, gives an authentic flavor, though regular mild or hot paprika can be used instead.

Ingredients

1 lb 2 oz (500 g) lean pork, diced

3½ oz (100 g) pork fat, diced

14 oz (400 g) good-quality sausage meat

1½ tbsp smoked paprika

3 garlic cloves, crushed

½ teaspoon freshly ground black pepper

1 oz (25 g) fine salt

1 Finely chop or mince the pork and fat depending on how coarse- or fine-textured you want your chorizo to be.

2 Put in a large bowl and add the diced pork and pork fat, paprika, garlic, pepper, and salt. Use your hands to thoroughly work the ingredients together so the flavors are evenly combined.

3 Prepare and fill the casings (see page 51).

Tips

For cooking chorizo, dry for just a week or two before adding to Mediterranean-style stews and casseroles or for flavoring paella.

Air-dried ham

Air-drying a whole joint of ham is one of the most rewarding but, it has to be said, most challenging of all the cured meats as the size of the joint makes it more susceptible to deterioration. The rewards of producing your own air-dried ham however, which can compete in flavor with expensive bought versions such as prosciutto and Serrano ham, make it well worth attempting at home. You need a little patience with this procedure as you won't see or taste the finished results for six to eight months. The ham is cured in two stages. First it is cured in salt under a heavy weight to extract as much moisture as possible, and then air-dried to mature and develop the flavor.

Protecting meat as it hangs

The high salt content of air-dried hams can discourage inquisitive flies and other pests but this cannot be guaranteed! For extra protection you can wrap the meat in a large piece of muslin (or a tube of muslin "stockinette"), sealing it at the ends with string. The open weave of muslin will allow the air to circulate. The other alternative is to make a protective "cage" for the meat to hang inside (see pages 57–58). This is also useful for drying other home cures such as salami and chorizo.

Check the condition of the pork frequently as it cures. If it starts to seep juice or become smelly (as opposed to smelling of appetizing deli meats) the meat is no doubt turning rotten and needs to be thrown away. At this stage there are no remedies and it is simply not worth the risk of proceeding. Hopefully this will not be the case and within five to eight months, the meat will be ready to eat. Test by squeezing the meat firmly. It should feel firm, but not completely rock-hard. You might find an area of light mold. This can be removed with a cloth dipped in vinegar, or cut away altogether. Slice the ham as thinly as possible, removing wafer-thin shards with a sharp knife.

How to build a cage for air drying

Building a cage for air drying meats is a failproof way of keeping flies and other unwanted bugs away. There are many ways of building one, the basic idea being to create a small structure completely enclosed in wire mesh or muslin, which must not have any gaps or holes. It can be as professional-looking or "botched" as you wish—neither will be more effective than the other as long as the air can move freely in and through it. The size required depends on what you are drying. If you are making one for a few small salami then the cage needn't be more than 12 in (30 cm) in width and a similar depth. If hanging a whole ham it will require a depth of at least 18 in (45 cm).

There are various items that could easily be converted into a suitable container. Cheese boxes (or larders as they're sometimes called) with wooden frames could easily be used by removing any shelving as well as the solid back, if there is one, replacing it with a mesh panel or layer of muslin. The box could be turned on its side for hanging and fitted with a hook or hooks inside.

A rummage in the shed or garage might unearth other suitable frames that you could easily convert. Bear in mind that you don't want anything too heavy (as it'll be much heavier again once the meat is inside) or awkward to get into for checking your meat's readiness.

If you are confident with woodwork, a frame could be built quickly using metal brackets (one in each corner) and lengths of wood, securing the wood at the corner brackets with screws. This could then be fitted with wire around the top for hanging and then covered with muslin or muslin stockinette and tied with string at the base so that it can be untied for checking progress (see Fig. 1).

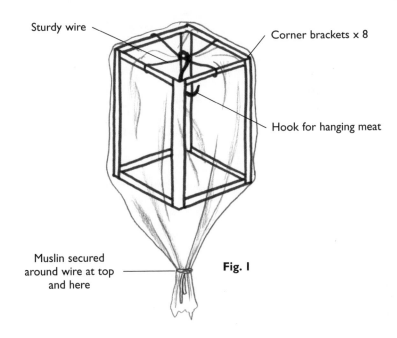

Sturdy wire

Corner brackets x 8

Hook for hanging meat

Muslin secured
around wire at top
and here

Fig. 1

Alternatively take two sturdy coat hangers, cross them over to create a square frame and secure with heavy duty tape, both at the necks of the hangers and where the horizontal bases of the hangers cross over. Fit a length of muslin stockinette between 20–32 in (50–80 cm) over the coat hangers. Bend two or three further coat hangers into loops and secure with tape. Push the loops of coat hanger up inside the muslin at 8 in (20 cm) intervals and secure these in place by sewing with a needle and thread in several places or by securing with lengths of floral wire.

Once the meat has been hooked up inside the frame, tie the muslin ends with string both around the top and base (see Fig. 2). If hanging salami or chorizo, use the horizontal wires of the coat hangers for stringing the sausages to. For a larger piece of air drying ham, the coat hangers won't be strong enough to support the meat so use two large meat hooks linked together, one for the meat and one for hanging.

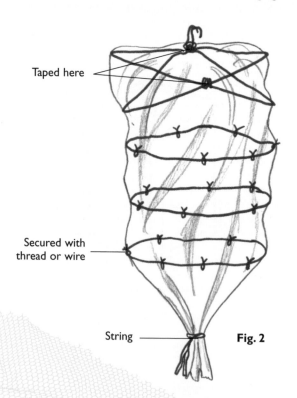

Taped here

Secured with
thread or wire

String

Fig. 2

Air-dried ham

Plan ahead by ordering a good-quality leg of pork and ask the butcher to tunnel bone it for you (remove the bone while keeping the leg intact). Once the curing is finished, the ham can be left hanging in a cool place for several months more so you can cut wafer-thin slices off as you please.

Ingredients

1 leg of pork, tunnel boned

Approx 17 lb 6 oz (8 kg) fine salt

1 Weigh the pork and lay it skin side down on the work surface. Check for any areas of blood or veins and remove. Rub plenty of salt through the center of the meat, making sure you spread it into all the folds and crevices of the joint.

2 Spread a ¾ in (2 cm) layer of salt into a large container, such as a wine case or large plastic storage container. Position the meat over the salt, skin side up, and cover the joint with the rest of the salt so it is completely submerged.

3 Cover with a piece of wood or plastic that fits inside the box. A small tray, board, or lid from a smaller storage container is ideal. Position a heavy weight, about one and a half times the weight of the pork, on top. A stone or piece of rock is ideal.

4 Calculate the curing time. For every 2 lb 4 oz (1 kg) of pork, allow 3 days curing (and make a note in the diary!).

5 Lift the pork from the salt and wipe off the excess with a damp cloth. Push a thoroughly clean meat hook through the meat and hang in a suitable place to dry. This process will take at least five, more likely six to eight months.

Biltong

Biltong is a South African recipe for salted and dried meat. Cured in small strips, it is easy to make at home as the whole process takes just a couple of days. Lean cuts of red meat such as beef, venison and ostrich make good biltong. In warm climates, it is traditionally air-dried but a fan-assisted oven makes a great substitute when the weather is not so warm and dry.

Ingredients

2 lb 4 oz (1 kg) piece lean beef such as topside, sirloin or fillet

2 tbsp fine salt

Cider or white wine vinegar

4 tbsp coriander seeds, finely crushed

1 tbsp freshly ground black pepper

4 tbsp light muscovado sugar

1 Cut the meat across the grain into ¾ in (2 cm) slices, then cut the slices into strips about ½ in (1 cm) wide. Ideally the pieces should be about 6 in (15 cm) long but this will depend on the cut of the meat.

2 Sprinkle the strips with the salt and place in a colander over a plate. Leave to stand for 2–3 hours until a pool of liquid from the beef has dripped onto the plate. Pat the meat dry on paper towels and transfer to a plate.

3 Use a pastry brush to brush each strip with the vinegar. Mix together the coriander, pepper, and sugar and place on a plate. Roll the pieces of beef in the spice mixture and space the pieces slightly apart on a tray. Leave to marinate overnight in the fridge.

4 Preheat the oven to its lowest setting. The temperature should be no higher than 225°F/110°C. Space the meat strips slightly apart on a wire rack and dry out in the oven for 4–5 hours, by which time they will have lost about 30–40% of their initial weight. The meat is ready when it is difficult to bend but is not yet brittle. Remove from the oven, leave until completely cold, and place in an airtight container. Store in a cool place for up to four weeks.

How to use

Biltong was traditionally made as nourishment that would not spoil during hot weather in the days before refigeration. It is still primarily used as a snack rather than a recipe ingredient. Excess quantities can be frozen for longer term storage.

Tips

• Venison also makes good biltong. Use a lean, tender cut such as haunch or loin.
• If you have time, put the meat in the freezer for a couple of hours to firm up a little and make slicing easier.
• Once dried, the biltong should be dark in color and feel firm on the outside but give a little when gently pressed.

Air-dried salted beef

This recipe imitates bresaola, the Italian cured meat that is delicious served simply in wafer-thin slices with a drizzle of olive oil and squeeze of lemon juice. When sliced, the meat will be brown around the edges but paler and softer in the center.

1 Trim off all the surface fat from the meat and place in a non-metallic container in which the joint fits fairly snugly. This is important, as the joint will need to be completely immersed in the marinade.

Ingredients

4 lb 8 oz (2 kg) piece beef top rump

1 oz (25 g) rosemary sprigs

3 bay leaves

2 tsp black peppercorns

1 tsp crushed dried chillies

4 garlic cloves, crushed

10½ oz (300 g) fine salt

7 oz (200 g) caster sugar

1 bottle Italian red wine

2 Pull the leaves from the rosemary sprigs and put in the small bowl of a food processor, or a coffee grinder reserved for grinding spices, with the bay leaves, peppercorns, and dried chilli and grind almost to a powder. Tip into a bowl, add the garlic, salt and sugar and mix well.

3 Rub the mixture all over the surface of the beef, turning the meat so it is thoroughly coated. Cover loosely with plastic wrap and chill for 24 hours.

4 Pour the wine into the dish and marinate for a further week, turning the meat occasionally if it is not completely submerged.

5 Remove from the marinade and drain thoroughly, patting dry with paper towels. Wrap in muslin, tying securely at both ends with string. Tie further lengths of string around the beef at 2 in (5 cm) intervals. This will help the beef to keep its shape as it hangs.

6 Hang in a cool, dry place for 2–4 weeks, placing a bowl or plate underneath to catch any juices. The meat will be ready when firm to the touch.

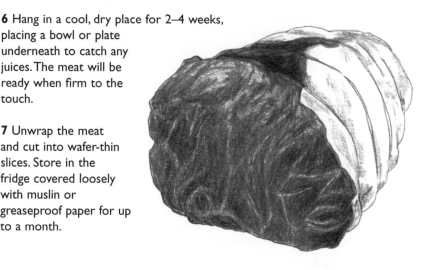

7 Unwrap the meat and cut into wafer-thin slices. Store in the fridge covered loosely with muslin or greaseproof paper for up to a month.

How to use

Air-dried beef can be served in numerous ways but keep it simple. Shave off small pieces and scatter over leafy salads with Parmesan shavings and a drizzle of olive oil. It's also delicious with wedges of sweet, juicy melon or scattered over a simple pizza.

Confits

Confits are a way of preserving by cooking and sealing meat in fat, usually its own, which inhibits the growth of bacteria. The meat is marinated in a salty rub before cooking to intensify the flavor and draw out moisture. Goose and duck are most frequently used because of their naturally high fat content, which renders down during slow cooking and contributes to the sealing fat. The secret of success is long, gentle cooking so the meat is meltingly tender and falls from the bone in soft, juicy threads. Encased in its protective fat, your finished confit can be stored in a cool place for several months before use. If you don't have a suitable cellar, larder, or other very cool place that will keep the sealing fat solid, use the fridge.

Storage containers

After cooking, you'll need to pack the meat into a container in which it fits quite snugly. Kilner jars are ideal. These are available in several sizes and come with a clip-on lid and rubber seal, or a metal lid and screw on metal ring. Thoroughly wash the jars before use and put in a low oven, 300°F/150°C for about 15 minutes before filling. Other suitable containers are ceramic or glazed earthenware terrines with lids. These are best stored in the fridge as their seal will not be completely airtight.

Using confits

Confits are used mainly as an ingredient in dishes such as cassoulet or cooked with beans or lentils for a delicious, wintry treat. Fruity sauces like those you'd serve with game make lovely accompaniments, balancing the richness of the meat. For summer eating, shred the crisped meat into salads or serve with warmed bread, pickled onions, gherkins, chunky apple sauce, or roasted shallots.

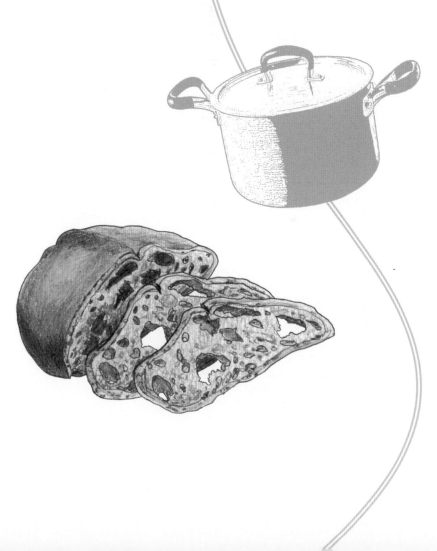

Duck confit

Duck confit is absolutely delicious, with meltingly tender flesh and golden, crispy skin. Perfect served simply with vegetables or used in a cassoulet or stew, you can vary the herbs and seasoning to suit your own taste.

Ingredients

6 large duck legs

1 oz (25 g) sea salt

1 tbsp finely chopped thyme

2 bay leaves, crumbled

6 garlic cloves, crushed

1 lb 7 oz (650 g) duck or goose fat

1 tsp freshly ground black pepper

1 Place a duck leg skin side down on a chopping board and cut through the joint. Scrape away the meat from the thigh half of the bone and remove. Repeat with the remaining duck legs. (This step isn't essential but it means there is less bone and the legs can be packed more tightly for cooking and sealing in the containers).

2 Mix together the salt, thyme, bay leaves, and garlic and rub all over the surfaces of the duck legs. Place in a shallow, non-metallic container. Cover loosely with plastic wrap and chill for 24–48 hours.

3 Preheat the oven to 300°F/150°C. Lift the duck from the dish, scraping the marinade back into the dish and reserving. Pat the meat dry on paper towels. Heat two tablespoons of the fat in a large frying pan and fry the duck legs, skin sides down, until golden. Place the legs into a flameproof casserole dish in which they fit quite snugly.

Add the reserved salt mixture and fat to the pan and heat gently until the fat has melted. Cover the legs with the fat mixture and cook in the oven for 1½–2 hours or until the meat is so tender that it can easily be pulled from the bone. Leave until cool but not set.

4 Lift the duck from the fat and pack into containers so they fit quite snugly. Cover with the fat, making sure all the meat is completely submerged. Cover with lids and store in a cool place.

5 To serve, preheat the oven to 425°F/220°C. Remove the duck from the fat, scraping off the excess, and place in a roasting tin, fat sides up. Bake for 20 minutes or until hot all the way through and crisped on the outside.

Using other meats

Use the same method for goose legs or whole, jointed goose or duck. It's also a great way of adding additional interest to chicken legs or guinea fowl. For a pork confit, use a mixture of cubed belly pork and lean leg of pork.

Re-using cooking fat

Don't discard the fat once you've used the meat. You can melt it down in a pan, strain through a sieve and store in the fridge for several months. It is delicious dotted over potatoes for roasting.

Rillettes

French rillettes use the same technique as confit making, but the meat used is cut into smaller pieces and shredded, giving a more pâté-like consistency. Use a fairly lean piece of belly pork or combine a mixture of belly and lean shoulder for perfectly tender, delicious results. Sealed under a layer of fat, rillettes can be stored in the fridge for several months and is best served simply with warmed baguettes. Once you've started using it, it'll keep for a further few days.

Ingredients

2 lb 4 oz (1 kg) lean belly pork, skin removed

4 oz (125 g) duck or goose fat

2 tbsp chopped thyme

Finely grated zest of 1 lemon

8 juniper berries, crushed

Salt and freshly ground black pepper

1 Preheat the oven to 275°F/140°C. Chop the pork into small pieces and place in a shallow, ovenproof dish or roasting tin. Scatter with the thyme, lemon zest, juniper berries, and a little salt and pepper.

2 Cover with a lid or foil and bake in the oven for 3–4 hours or until the meat is meltingly tender and lightly browned.

3 Leave the meat until cool enough to handle then drain off the fat.
Tear the meat into shreds using two forks, breaking it up into very small
pieces. Tip into a bowl and beat in a little salt and plenty of black pepper.

4 Pack the meat firmly into clip-top or screw-top jars.

5 Scrape the reserved cooking juices into a small saucepan and heat until
liquid. Strain through a sieve into a jug and pour a thin layer over the meat
until just covered. Tap the containers on the work surface to help displace
any air bubbles and add a little more of the fat if necessary. Cover with the
lids and refrigerate until ready to serve.

Smoking

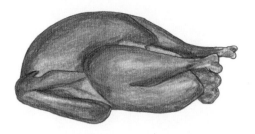

Smoking is about celebrating food, putting on a show, thoughtful planning, and a bit of imagination. Since time began we have smoked our food to protect it and keep it safe to eat. Now we smoke food because it makes it taste remarkable. Smoking also offers something a little bit different. We've all been to barbecues, but to be entertained by someone who's just transported a whole smoked turkey or tray of mussels from smoker to table is rather more impressive—and such a talking point!

The three principle processes associated with smoking food are flavoring, cooking, and preserving. The first step is to decide what you want to achieve, which is usually determined by what and when you want to eat. When planning your smoking sessions you'll find it really useful to "countdown backwards" to allow time for marinades and rubs to be absorbed, pre-brining and drying to take place, to get the fire going, the cooking itself, and of course the finishing touches. Your second decision is whether to blow hot or cold.

Cold smoking

Cold smoking does not cook food but imparts the special flavors associated with the woods, spices, and herbs and brines that you want to use. Remember that the smoke is absorbed slowly so the longer the smoking time, the stronger the taste. You can over-do it! After cold smoking you can finish cooking on a barbecue or fire pit, or of course you can always go indoors and use the oven or store the cold-smoked food in the fridge for later on. Temperatures for cold smoking are between 50°F (10°C) and 82°F (28°C). At this low temperature food will take on flavor but should remain moist. Some cold smoking can be done in an hour, sometimes you may want to keep the process going for two or three days.

Hot smoking

Hot smoking is about smoke and heat and an element of control. If you get your timing right and keep the home fire burning there will be no need to re-heat or re-cook because hot smoking occurs within the range of 165°F (74°C) to 240°F (115°C). At these temperatures foods are fully cooked, still moist, and will have taken on all the flavor you could wish for. For comparison, the traditional barbecue operates at around 350°F (180°C), so hot smoking offers a greater element of control.

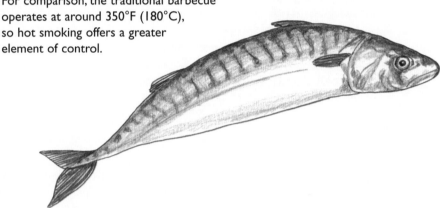

How it works

Smoke is an antimicrobial and antioxidant which penetrates the surface of food. In itself smoking is insufficient to preserve food for long, so curing techniques such as salting or drying are used before smoking. Drying, curing, and the other techniques set out in this book will inhibit bacterial growth by extracting moisture, which is what causes fats to break down and to "go off". Smoking also provides the exterior surfaces an extra layer of protection. Smoking is especially effective for preserving oily fish such as salmon, mackerel, and tuna, as its antioxidant properties protect the surface fat. Some heavily-salted, long-smoked fish can keep without refrigeration for months.

Fuels

You can use just about any non-toxic material that gives off smoke as fuel. You can smoke hams over burning corncobs or experiment with hay, seaweed, sawdust, peat, tea, and rice. Around the world there are some very unusual ways of smoking foods, such as the Icelandic method of smoking dried sheep dung to cold smoke whale meat! More conventional smoking methods usually combine the use of woods for providing smoke and flavor and charcoal as a propellant, keeping the energy and heat of the fire maintained. Of course you can buy prepared woodchips and sawdust from any supplier but it is worth making a collection for yourself. That way you will always have a choice of woods at hand and cuttings and prunings are a lot cheaper. Some barbecue fans cannot be torn away from charcoal brickettes of the molded, fuel-injected variety. For smoking, which takes longer than barbecuing, you really need to go for a good lumpwood charcoal as this burns more evenly and is a natural product to expose your food to.

It is helpful to understand how the wood imparts flavor and preserves food as we decide which to choose. Hardwoods comprise three main materials: cellulose and hemicellulose are sugars which provide color and fruitiness; lignin provides the smokey and roasted flavors and a distinctive vanilla sweetness. Some softwoods, especially pines and firs, also contain resin which produces an acid-tasting soot, so while these might be good to get the campfire started, avoid using softwoods for smoking.

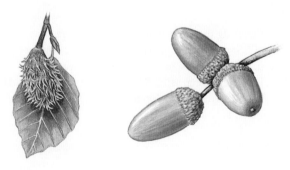

Bear in mind also that some trees are poisonous, at least in part. Toxins are usually found in berries, leaves, and fruits but it would be best to avoid rhododendron, horse chestnut, elder, holly, laurel, or yew. This still leaves you a lot of very good wood choices. In Europe, alder, beech and oak are the traditional smoking woods along with fruit woods. In America, hickory and pecan wood are popular for their tanginess.

Each tree has its own balance of sugars and lignin, and so each flavors food in its own way. They also burn at different temperatures. Denser woods such as oak tend to burn hot so, to keep them smoldering, we need to adjust air flow and moisture. The choice of wood is entirely up to you but here are some that will serve well. Experiment!

Fuels		
Wood	**Characteristics**	**Suitable for**
Hickory	Strong, "meaty" aroma	Pork, chicken, beef, game, sausages
Alder	Delicate, a flavor that enhances	Lighter meats and fish, cheeses
Oak	Strong, pleasing, earthy aroma	All meats, smoked salmon sausages, cheese, nuts
Beech	Mild, similar to oak	All meats, shellfish, cheese
Maple	Mild, slightly sweet	Ham or bacon, vegetables
Cherry	Mild, slightly sweet	Poultry, game birds, pork
Apple	Fairly strong, slightly sweet	Beef, poultry, game birds, pork
Peach or **Pear**	Slightly sweet, aromatic	Poultry, game birds, pork, cheese
Grape vines	Strong aroma of spice, and fruit	All meats, cheese

Gathering wood

As you gaze around your own garden and beyond into your neighbours' gardens, you may be disappointed to discover little more than the odd apple tree among the rhododendrons or some municipal beech tree under the close protection of the local authority. Unfortunately, these trees are not available to you as resources. So where do you go to get your wood? You could of course order it. There are many suppliers offering a great range of smokers' woods (see page 124), cut conveniently into manageable blocks, but a small bag is expensive, especially so for some sawdusts. The alternative is to collect your own when you see cheaper supplies, setting aside a dry storage area such as a garden shed, garage, or boiler room. In this way you can save money and source your own fuels.

A visit to a farm shop or farmers' market should, on the purchase of some good local fruit, offer the opportunity for you to ask when the orchards are being pruned, usually in the winter. Of course sawdust is something of a by-product at the average sawmill. It doesn't hurt to ask for some oak or beech wood dust. What you mustn't do is go foraging around the park at night with a chainsaw and an old sack!

Where to smoke

Think about where you are going to smoke. If you live in an apartment building where neighbors are likely to call the landlord if you so much as burn toast, then you may have to borrow a friend's garden to practice your smoking. Or, of course, you can head off into the wilderness armed with your bread bin and a box of matches. If you have a garden and you are allowed to barbecue then you are also allowed to smoke. Smokers get hot so don't start it against the shed, fence, or neighbor's wall—they will not thank you for offset smoking their curtains! Do remember that smoking is a longer process than barbecuing and although many love the smell of home-smoked ham, they might not love it every weekend. Also consider wind direction, open windows, and clotheslines, basically the same considerations as you would for barbecuing or any other form of fire lighting. Smokers are heavy, cumbrous, and not particularly easy to move. Think about an area in the garden where you can leave it semi-permanently, preferably somewhere that's sheltered from strong winds, heavy rain, and extreme temperature conditions that will have an effect on the internal temperature of the smoker. Wherever you set up your smoker, it is important to remember that you are playing with fire. Set the smoker well away from treated wood, gas canisters, or a shed full of old paint tins.

Your smoker will still work during a snow storm, freezing rain, or ice-cold weather conditions, but you'll need to be that bit more vigilant and keep the fuel topped up. In the same way, if it is hot weather, you will need to ensure that the vents are open and the water bowl, if you are water smoking, is topped up.

What to smoke and when

One of the advantages of smoking is that you can make the most of seasonal produce but you are by no means restricted by it. The strength of the smoke flavors will always bring out the best even from foods that have been frozen up to the limit of their recommended freezer life, giving them a more interesting flavor that they might otherwise lack.

Seasonal produce for smoking
Spring
Pigeon, asparagus Mackerel, halibut, bass, sardines Welsh lamb Crabs, salmon trout, lobster, and shrimp
Summer
Clams, squid Sweetcorn, courgettes Grey mullet, haddock, herring, bream, pollack, and red mullet
Autumn
Mussels Rabbit, partridge, grouse, guinea fowl, hare, and mallard Venison Wild nuts, onions, peppers, pumpkin
Winter
Brill, flounder, whiting, oysters, scallops, wild salmon Goose, turkey, pheasant Conger eel Rabbit and hare

Non-seasonal foods

Bacon, ham, beef, pork, sausages and offal, chicken, farmed salmon, cod, and quail can all of course be smoked at any time of the year. Other suitable ingredients include clams, garlic, eggs, and cheese. In fact, almost any food can be smoked and there is a lot of fun to be had in experimentation.

Before you start

Smoking, both hot and cold, requires a little more planning than a regular barbecue as some foods will need curing and drying beforehand or simply rubbing with spices or other flavorings. This might simply take an hour or so in the case of smoked salmon but several days for a piece of bacon. Decide on exactly what you want to smoke before you start.

Brining foods for smoking

Some foods are brined before hot or cold smoking. The cure draws out the excess moisture from the food, adds flavor, and inhibits the growth of bacteria. It also extracts proteins from the food which settle on the surface and give smoked foods their familiar sheen. After curing, foods should be dried properly before smoking, either by hanging from a hook in the fridge, or on a wire rack in a cool place for several hours until no longer wet-looking. Do not let the food dry out too much, however, or a skin will form that will not let the smoke penetrate through the surface.

Equipment

Some of the equipment required for smoking will be the same as you've used for curing ingredients, e.g., kitchen scales, containers, meat hooks, muslin, etc. Here are some additional useful pieces of equipment, besides of course the smoker itself.

- A **thermometer** will help you to monitor the temperature inside the smoking chamber, unless of course you've bought a smoker with an integral one.
- **Long matches** which will be easier to maneuver inside the smoker than regular kitchen ones.
- A **meat thermometer** for checking the internal temperature of the meat after smoking.
- Long handled **tongs** are useful for arranging foods inside the smoker.
- Metal framed smokers can get very hot. **Oven gloves** are useful for opening door handles and rearranging racks.

Tips

• If you are unsure of whether the food is sufficiently smoked, cut a little off and do a taste test. Remember that if you are cold smoking the food it won't be cooked so fry your taste taste lightly in a pan to check that you are happy with it before removing the whole lot from the smoker.

• Don't try to smoke using wet fuels. If your sawdust is overly damp, lay it out in a container and leave in a warm place such as the airing cupboard, boiler room, or in a low or cooling oven after cooking. Don't use a hot oven or the sawdust could catch fire.

• Clean out the smoker after each session so that ash does not accumulate.

• Some cold smoked foods will continue to develop in flavor once removed from the smoker. Ideally they're best left for a day before eating.

• If you've cured, marinated, or prepared a rub for your meat or fish, make sure it's prepared and ready to go before handling the charcoal.

• Don't forget to weigh the food before cooking large pieces of meat as weight determines cooking time.

• Always rinse brined meats and fish under cold running water, then dry thoroughly on paper towels before. Wet foods won't smoke very well.

• Before chilling, wrap smoked foods in foil or greaseproof paper so their flavors do not penetrate other foods.

Smokers

In its crudest form, smoking can be achieved by hanging fish or meat above an outdoor fire or by positioning it somewhere above the flames of an indoor fireplace. In both cases the food needs to be adequately away from the flames so it's not burnt to a cinder! Using smokers is a far more reliable way of smoking food. Some smokers are specifically designed for hot or cold smoking while others can be used for both by controlling the amount of heat and smoke generated. There is an increasingly wide range of smokers available through various sources, particularly internet suppliers (see page 124). Experienced smokers might have a substantial, walk-in smokehouse set somewhere on their grounds or build a smaller, fixed structure using brick, wood, or metal frames. For first-time smoking it's best to start with a small, less expensive structure and instructions for building your own are set out below. Once you have explored whether you are cut out for this style of self-sufficiency, you might want to splash out and buy a more substantial smoker, maybe one day investing in that top of the range fixed structure. Do bear in mind that a substantially sized smokehouse might need zoning permission. On a much smaller scale, smokers can loosely be divided into four main types:

Offset smokers

Drum smokers

Water smokers

Box smokers

Offset smokers

The main characteristic of the offset smoker is that the cooking chamber and firebox are separate. The heat and smoke from the fire are drawn through a connecting pipe or opening into the cooking chamber. The longer the connecting pipe, the cooler the smoke will be. Most manufacturers' models are based on a simple but effective two-chamber design. Offset smokers are always used for cold smoking.

Drum smokers

An upright steel drum smoker can be used for hot smoking and also provides for most offset smoking as the heat source is a few feet below the cooking racks. Temperature is controlled by limiting the amount of air intake at the bottom of the drum and the rate you allow the smoke to escape. It is, however, harder to control the temperature and smoke supply as the fuel is contained within the same container.

Water smokers

A water smoker is a variation of the drum smoker that can be used reliably for hot or cold smoking. It uses charcoal and wood to generate smoke and heat, and contains a water bowl between the fire and the cooking racks. The water bowl provides humidity, which helps keep foods moist and also keeps the temperature constant. The water bowl also catches fat or oil that prevents flaring.

Box smokers

These are small metal smokers not much larger than a bread bin and used for hot smoking. In its simplest form, a box smoker is a steel box with a lid, a base for scattering the wood or sawdust, and a tray for resting the smoking foods on. If you want to experiment with home smoking but don't want to build your own, this is a good smoker to buy, as they are reasonably inexpensive. However you could try building the simple box-style smoker on page 88.

Using a barbecue for smoking

If you have a kettle or drum barbecue, or any other barbecue with a lid, you can use it to smoke foods, provided it has an air vent to control the air flow. Use charcoal in the same way as you would for barbecuing but position it either side of the barbecue base. Add sawdust and dampened wood to this so the heat intensity is reduced. The food is then cooked on the rack over the center of the barbecue so that it is not directly over the coals. If the coals start to flare up, close the air vent so they are reduced to a gentle smolder.

Building your own smoker

If you are not inclined to rummage for materials, nor are you in any way adept at assembling them even if you rummaged successfully, then a reasonable investment in a commercial smoker will keep you smoking for years. But it is worth experimenting with making your own, at the same time saving money that is so much better spent on the food itself. By building a simple drum smoker you will be able to hot smoke or convert it into a water smoker. With a little adaptation it will also serve very well as a cold smoker.

Equipment

For a hot smoker:
- Wooden barrel or steel drum
- Drill and bits
- Saw
- 4 x 4 in (10 cm) bolts, washers and nuts
- Picnic barbecue tray or disposable foil barbecue base
- Bricks for supporting smoker
- Barbecue rack or racks
- Small empty food can

Additional equipment for an offset (cold) smoker:
- Old stove
- Length of domestic quality aluminium silver ducting hose

Additional equipment for a water smoker:
- Water bowl
- 3 x 6 in (15 cm) bolts

Drum smoker

The smoke needs to be captured in some form of box, whether metal or wooden. If you have tools to cut and drill metal or heavy wood, then a steel drum or barrel is the perfect starting point.

The easiest way to go about drum smoking is to convert a steel drum with an old barbecue basket to hold charcoal near the bottom and then to secure cooking racks near the top and cover with a vented lid. A 45-gallon oil drum can be bought from all good agricultural hardware suppliers or you might be lucky enough to get one online. A wooden barrel will set you back a bit more but once you've got it and adapted it, you'll no doubt find that it will last for years.

Home-made drum smoker—set up for cold smoking

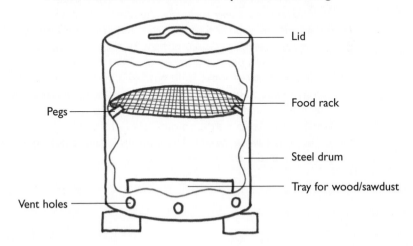

Lid

Food rack

Pegs

Steel drum

Tray for wood/sawdust

Vent holes

As the wood, whether dry or dampened, gets hot it generates smoke. A fuel tray is needed for the charcoal and wood or sawdust. A small, portable, metal picnic barbecue tray is idea and can be raised on its legs, if it has them, or simply placed on bricks. Alternatively you could use the foil base of a disposable barbecue and raise the base on bricks. You don't need fire bricks; an ordinary brick will withstand temperatures of 2192°F (1200°C), far hotter than anything you will be cooking.

The drum or barrel needs to be separated from its base in order to safely start the fire going, and also separated from its top in order to put the food onto racks. Saw out the base and rest this on raised bricks to create a sturdy foundation for the smoker. If the drum or barrel has no lid, cut out the top in the same way so you are able to arrange, and re-arrange your foods.

The fire will also need to draw air and therefore you will need to drill holes around the lower end of the container below the height of the fire. Similarly you will need to improvise an adjustable vent at the top in order to allow the smoke to pass out of the smoke chamber or the fire will choke. Cut out a small hole, about 2 in (5 cm) in diameter from the lid. Have a small, empty food tin (baked beans or tomatoes!) for positioning over the hole.

The easiest way to secure the food racks is to drill three holes around the top third of the drum and insert six-inch bolts from the outside inwards and secure them from the inside with washers and bolts. This will provide a strong platform for the racks to rest on. You can buy barbecue racks to fit but, again, you could make them from scratch with stout wire frame and by cutting heavy-duty mesh to fit. You can also add another rack about 6 in (15 cm) below this one so you can double up on the amount of food smoked.

You now have a smoker that is set up for hot smoking. It can easily be converted into a cold or water smoker.

How to convert to an offset (cold) smoker

Your homemade hot smoker can easily be adapted for cold smoking. You will need to drill a hole wide enough to take ducting from a heat box into the smoking chamber you have built. There are various items that you can use as a heat box. A search of agricultural suppliers' websites or a trip to the recycling center should reveal a metal box strong enough to act as a temporary oven. An old stove on legs is ideal. Remember that any metal containers you use, whether as a smoke chamber or the firebox, should never be made of or treated with zinc or any other toxic metals. Look for steel or iron. You will also need domestic-standard flexible piping. A length of aluminium silver ducting hose from a DIY store is ideal. Cut a hole of the same diameter as the hose in the side of the drum (or barrel). If the hole is cut fairly accurately, the piping should sit comfortably inside the hole. For convenience, cut the piping to about 6 feet long. This will give the smoke time to cool before it reaches the smoking chamber.

Home-made offset smoker cold-smoking

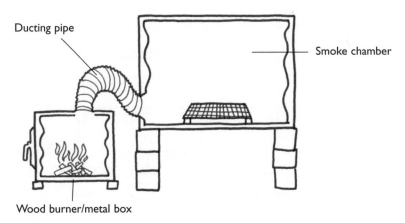

Ducting pipe

Smoke chamber

Wood burner/metal box

...and to a water smoker

The advantage of a water smoker is its ability to maintain a more constant temperature. To convert your drum smoker to a water smoker, take three more 6 in (15 cm) bolts and drill three more holes about 18 in (45 cm) above the highest point of the fire source. You will now need a metal bowl, the sort of thing a large dog eats from, which you can fill with warm water once the fire is ready to start cooking. Resting the water bowl on the bolts above the fire generates some steam with the smoke and helps to keep the food and fire separate. Of course, the smoke will circulate around the sides of the bowl so you don't want a tight fit between the rim and the walls of your smoker.

Portable box/picnic smoker

Small box smokers are relatively inexpensive to buy but you can still have fun making your own cheaper version. This style of smoker is sometimes referred to as a fisherman's smoker, as fisherman would take them angling, ready to smoke their catch.

Home-made box/picnic smoker for hot-smoking

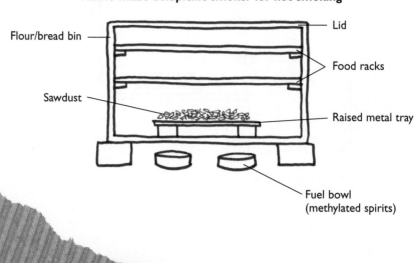

Flour/bread bin

Lid

Food racks

Sawdust

Raised metal tray

Fuel bowl
(methylated spirits)

Equipment

- Metal container such as a flour or bread bin
- 4 metal corner brackets and screws or bolts and washers
- Cooking rack
- Metal tray
- 4 stones or metal supports
- Bricks for supporting smoker
- Metal containers for methylated spirits

The easiest way to make a portable smoker is to find a metal flour or bread bin, or any other similar shaped item, that has a lid. Check out flea markets, charity stores and the internet. Make a shelf about a third of the way down the bin by making four supports—two on each side. Use either metal corner brackets or four small bolts so that you can rest a barbecue rack on them. A wire cooling rack, barbecue rack, grill rack, or small oven shelf is ideal. Place a metal tray in the base (such as an old baking tray) raising it up by about 1 in (2.5 cm) on anything that can withstand the heat. Have a rummage around the shed, garden, or DIY store for these (garden stones, metal blocks, or brackets are all suitable). Place a pile of sawdust on the tray. Raise the box onto bricks and beneath it light one or two metal containers filled with methylated spirits. Do not substitute with any other fuel, it has to be meths, which burns at the right temperature, quite slowly. Once the wood starts smoking, position the food and lid. When the methhylated spirits runs out, the food is usually cooked; if it needs longer simply refill the meths and continue cooking.

Cold smoking

Cold smoking is a subtle and slow process that may easily last overnight or even up to a week. The longer the food is left in the smoker, the stronger the smoky flavor will become—this is a matter of personal taste. Because of the lengthy smoking time, it is possible that the wood shavings will go out somewhere along the line. If this happens, simply re-start the process and keep topping up the sawdust with a few more shavings. Don't worry if the smoker stops smoking altogether, for example during the night, as you can restart it in the morning. Always remember that cold smoking does not cook food (in the vast majority of cases), so your smoked products will require further cooking. Because of the low smoking temperatures, it is best to avoid cold smoking during hot summer months as you might not be able to keep the temperatures low enough. Refer to pages **84–87** for types of cold smokers and how to build your own.

Smoking times and temperatures

Smoking times are not as critical as they are for hot smoking. The

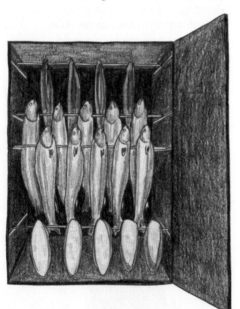

inevitable fluctuating amounts of smoke generated by your homemade smoker will also affect smoking times. A piece of cheese might take a couple of hours whereas a large piece of bacon could take several days. It is a case of regularly checking the smoker to see if you are happy with its progress. If you are unsure, you can always snip off a small piece and taste test it to see whether you are happy with its smokiness, but remember to cook it first. The temperature for cold smoking should hover between 50–82°F (10–28°C). It is crucial that the temperature does not go significantly above this or you'll be smoking your food somewhere between cold and hot smoking. This is when bacteria are more likely to multiply.

Drying before smoking
Wet or dry brined meat and fish must be dried before smoking, as you cannot smoke wet foods successfully. It also allows some of the proteins that are extracted by salting to settle on the surface and form a glazed film, which looks attractive and helps preserve the fish or meat. First pat dry lightly on paper towels then dry further by placing the fish or meat on a wire cooling and leaving for several hours. This can be done at room temperature if the weather is not too hot, or in the fridge.

Starting the smoker

Equipment

- Lumpwood charcoal
- 2 pieces of wood the size of doorstops, cut into shavings
- Several handfuls of sawdust
- Matches
- Firelighters or kindling

Unlike barbecuing or hot smoking, you need not wait for the flames to subside before adding the foods as the heat source is in a separate chamber and the smoke will have cooled before it enters the smoker chamber.

Start by placing the charcoal in the heat box and lighting with firelighters or kindling. Once the charcoal is glowing, surround the coals with sawdust and loosely add the wood shavings. The sawdust will start to burn slowly and catch the shavings as it goes. Once gently smoking, position the food on the rack or racks. Leave to smoke, checking that a constant supply of smoke is entering the smoker. The smoke should feel very slightly warm on your hands as it enters the smoker. If it feels hot you will need to lengthen the ducting hose that connects the two chambers. If the smoke subsides, check the heat box and top up with more wood shavings and sawdust.

Foods for cold smoking

Whether you hot or cold smoke food is entirely up to you and you can experiment with both types to see what you prefer. Some foods, like mackerel and salmon, work brilliantly both ways. Don't forget that brined meats and fish will need to be thoroughly dried before smoking. Here are some foods that are particularly good cold smoked. Whole fish will need to be cleaned and gutted before smoking. After smoking you can cook the food immediately, although keeping it in the fridge for another 24 hours often accentuates the smoky flavor.

Smoked Mackerel
Smoke mackerel whole or filleted. Brine cure (see page 40), dry and smoke for 6–8 hours or overnight.
Smoked White Fish e.g., cod, haddock, pollack, whiting
Use filleted fish and check over for any stray bones. Brine cure (see page 40), dry, and smoke overnight or for up to 2–3 days.
Smoked kippers
Use really fresh gutted herrings and cut off the heads. Open out the undersides of the fish on a chopping board so that the skin side is up. Press down firmly along the backbone with your thumb so the fish becomes completely flattened. Brine cure (see page 40), dry, and smoke for 6–12 hours.
Smoked Trout
Use large trout fillets and brine cure (see page 40). Dry and smoke overnight or for 2–3 days.
Smoked bacon
Use your home-cured bacon (either dry- or brine-cured). Dry and smoke to your taste. This could be 12 hours or several days.

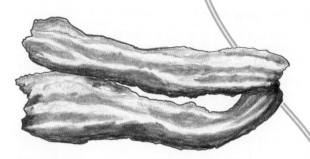

Smoked ham
Add further flavor to your home-cured ham. Dry and smoke for 3–6 days
prior to cooking.

Smoked cod roe
Brine cure two cod's roe, (see page 40), for 40 minutes. Carefully drain and
pat dry on paper towels. Dry and wrap loosely in muslin before smoking
for 12–24 hours.

Smoked eggs
Hard boil, peel, and cool however many eggs you would like to smoke.
Season with salt and pepper and smoke for 6–8 hours.

Smoked cheese
Most cheeses can be smoked, though some take up the smoky flavor
better than others. Cut cheese into small wedges or slices up to about
2 in (5 cm) thick before smoking. Firmer cheeses, such as cheddar and
Gruyere, can rest on a rack, while softer ones, such as feta and mozzarella,
are best smoked on a muslin cloth. Blue-veined cheeses do not smoke
well.

Smoked nuts
Most nuts smoke well, though almonds are particularly good. Use
unblanched nuts, scatter them on a fine rack lined with muslin to stop
them slipping through, sprinkle with salt, and smoke for 3–4 hours.

Smoked salmon

Salmon can be smoked as for smoked trout (see page 92). The cure in this recipe adds extra flavor and illustrates how you can develop your own cures. By experimenting with ingredients you can produce your own very unique artisan smoked fish.

Ingredients

9¾ oz (275 g) fine salt

5 oz (150 g) golden caster sugar

3 hot red chillies, thinly sliced

3 oz (75 g) fresh ginger, sliced

2 1 lb 2 oz (500 g) fillets of wild or farmed salmon

1 Mix together the salt, sugar, chillies, and ginger in a non-metallic container large enough to contain the fish without bending the fillets. A rectangular plastic storage container is ideal though you can use glass or glazed earthenware. Add 4 cups (1.2 litres) very cold water and stir well until the salt has completely dissolved.

2 Do the egg test (see page 33) to check that the brine is salty enough. Leave to stand for about an hour to allow the flavors to infuse.

3 Place the salmon in the brine and leave for 45 minutes. Drain the salmon and remove the excess water on paper towels. Place on a wire rack and dry for 4–6 hours.

4 Cold smoke for 24–48 hours, or to your preferred degree of smokiness.

Hot smoking

Hot smoking is preferred by the more spontaneous cook and those who want to dig in to hot food with a smoky flavor as soon as it's ready! It offers even more versatility than barbecuing as gentler cooking ensures that even large joints of meat can be safely cooked through. Allow at least 30 minutes to prepare the smoker before adding any foods. If at some point during smoking you need to open the smoker to add water or wood, you should allow an extra 15 minutes cooking time to let the temperature return to the required level.

Smoking times and temperatures

Hot smoking is an intuitive form of cooking but you can still control temperature by adjusting the airflow using the air vent in the top of the smoker. Air that is drawn in at the bottom needs to be vented out. By opening the aperture on the air vent (or moving your food can on your homemade smoker) you encourage a faster airflow, which increases the internal temperature. By closing the aperture the temperature is reduced. The various woods used in smoking burn at different temperatures. The heat from the fire should be smoldering and consistent, not flaring or choking. If the vent is closed and the fire looks like it is in danger of going out, open the aperture. If you have a raging inferno that will cook your foods to a crisp, reduce the airflow. This allows you to control the temperature within the range of 165–240°F (74–115°C), perfect for all hot-smoked foods.

As a general guide, you will need to allow one hour's smoking for each 1 lb 2 oz (500 g) of food in one piece. So an 11 lb (5 kg) turkey requires ten hours' smoking but two 4 lb 6 oz (2 kg) joints of beef, pork, or whole chickens will need four hours. Always allow at least 45 minutes as a minimum cooking time even for small fish or fillets that weigh less than a pound.

The trials of smoking

Like barbecuing, hot smoking is a fun and novel way to cook but, however much time and energy you put into keeping the fire burning and at an adequate temperature, there is no way that you can have the same control over a smoker as you can with a conventional cooker. In other words, there is an element of hit and miss. Many smoking books will give accurate temperatures at which foods should be smoked but the skill of temperature control comes with experience and practice, something that the novice home curer might never have the time to perfect. For this reason, the recipes in this book do not provide smoking temperatures for hot smoking, but a rough guide as to how long food will take. It is however essential that foods, particularly meat and fish, are cooked safely. It is well worth investing in a meat thermometer for checking the internal temperatures of large pieces or joints of meat and therefore whether they are sufficiently cooked in the center of the joint.

If you have hot smoked your joint of pork or whole chicken and the temperature has lowered to a point where it is no longer cooking the meat and there is insufficient time to reload the fire, you can easily put the food on a barbecue to finish it off or use a conventional oven. The food will still have a lovely smoky flavor so there is no need for any disappointment.

Using a meat thermometer

A meat thermometer is a great tool for checking that the internal temperature of the meat is ready for eating. Insert the point of the thermometer through to the thickest area of the meat (but away from the bone). Leave for 30 seconds until the gauge has registered.

Finished temperatures for meat

• Beef or lamb, rare	120°F (50°C)
• Beef or lamb, medium	140°F (60°C)
• Beef or lamb, well done	165°F (75°C)
• Pork	165°F (75°C)
• Chicken	165°F (75°C)
• Turkey	165°F (75°C)

Starting the smoker

Soak two of your pieces of wood in cold water for 30 minutes. Place the charcoal in the fire basket (or old metal barbecue tray, or disposable foil tray if you have made your own smoker) and light with firelighters or kindling, gradually building up the charcoal until the basket is full.

Wait for the charcoal to glow rather than burn. Add the remaining two pieces of wood to the basket. Position the food on the rack or racks and add the soaked pieces of wood to the basket.

Equipment

- 4 pieces of wood the size of doorstops
- Lumpwood charcoal
- Matches
- Firelighters or kindling
- Kettle of boiling water if using a water smoker

Note: If using a water smoker, position the water bowl once the fire has settled and fill with boiling water from the kettle immediately before you start to smoke

Wrapping foods for hot smoking

If you are cooking turkey or any light-colored meat that needs more than three hours exposed to smoke, wrap it in damp muslin or muslin stockinette before smoking. The muslin allows the smoke to reach the meat but prevents a build up of carbon or over-coloring. You can remove the muslin about three-quarters of the way through the cooking process and if you are happy with the color (which should be an appetizing burnished bronze), leave it unwrapped. If you do not want the turkey to brown further, leave it wrapped for the rest of the cooking time.

Foods for hot smoking

Hot smoking is more akin to barbecuing than cold smoking and you can flavor, season, and marinate your foods prior to smoking just as you might do on the barbecue. This is a chance for you to bring out your favorite recipes and try introducing a little smoke to them. Almost any foods can be hot smoked, including vegetables and fruits. Just consider what the flavor might taste like and if it appeals, then smoke it! Lamb doesn't respond to smoking as well as other meats but you could try a cheaper cut or a couple of lamb chops and see what you think. If you are serving foods freshly smoked, have any accompaniments and sauces ready as smoking times are quite short. Test whether the fish or meat is cooked through. Fish should flake easily when pierced with a knife and the thigh juices of a chicken should run clear when pierced with a skewer. Meats that are safe to eat rare when conventionally cooked, such as beef and venison, can equally be eaten rare when hot smoked.

Smoked Mackerel or Trout

Gut the fish and wash well under cold running water, cleaning out any blood from inside the cavity and leaving the heads intact. Pat dry on paper towels and brine cure (see page 40) for 1 hour. Dry and smoke for two hours. Mackerel is also delicious spread with a thin layer of grainy mustard before smoking and seasoned generously with black pepper.

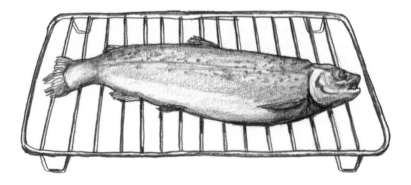

Smoked Salmon
Choose whole salmon fillets or cut the fillets into portions and brine cure (see page 40). Dry and smoke for two hours.

Smoked Sardines and Sprats
Only use very fresh fish and gut sardines (there is no need to gut sprats). Pat dry on paper towels, brush very lightly with olive oil, and sprinkle with a little chilli powder or hot paprika. Delicious served with a homemade spicy tomato sauce.

Smoked Chicken, Pheasant, and Guinea Fowl
Rinse out the bird and dry on paper towels. Season inside and out and pack the cavity with fresh herbs such as rosemary, thyme, parsley, fennel, coriander, oregano, or sage, or brush the skin lightly with olive oil and rub with a spice blend. Truss the bird to keep it compact and moist, weigh, and smoke for one hour per 1 lb 2 oz (500 g).

Smoked Pigeon, Quail, Partridge, and other small birds
Rinse and dry the birds and season inside and out with salt and pepper. Pack the cavities with garlic cloves and plenty of fresh herbs. Wrap thin cut bacon around the breast areas and secure with string. Smoke for 50 minutes to 1 hour, depending on size, or until the thighs feel tender when pierced with a knife.

Smoked Duck
Prick all over the underside of the bird with a skewer. Rub plenty of allspice, salt, and pepper into the skin. Truss with string to keep the bird compact and moist. Weigh and smoke for 1 hour per 1 lb 2oz (500 g).

Smoked Garlic
Use plump heads of garlic, leave whole and smoke for 1–1½ hours or until golden.

How to use smoked garlic
Smoked garlic can be used in any recipe where you'd use regular garlic, provided you want a slightly smoky flavor. It's particularly good in mayonnaise and sauces. Store in a plastic bag so its flavor does not taint other foods in the fridge.

Recipes

Now that you've made your cured meats and fish, you might be looking for some inspirational recipe ideas that will bring out the best in them. This is the purest form of cooking from scratch—not only are you cooking delicious dishes, you've even produced the main ingredients, which will be far superior to anything you'll buy.

Rubs and butters

If you are smoking a chicken or turkey, simple beef, pork, or fish steaks, a spicy rub pressed into the surface of the meat about 30 minutes before smoking will act as a dry marinade, infusing it with flavor. There are plenty of ready-made spice rubs available in supermarkets but you can just as easily mix your own from your spice collection. A flavored butter can add that finishing touch to smoked foods, spooned over before serving so it melts to a deliciously buttery sauce.

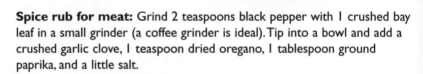

Spice rub for meat: Grind 2 teaspoons black pepper with 1 crushed bay leaf in a small grinder (a coffee grinder is ideal). Tip into a bowl and add a crushed garlic clove, 1 teaspoon dried oregano, 1 tablespoon ground paprika, and a little salt.

Spice rub for fish: Crush 2 teaspoons fennel seeds using a pestle and mortar. Tip into a bowl and add the finely grated rind of 2 lemons, ½ teaspoon celery salt, and plenty of freshly ground pepper.

Chilli spice butter: Crush 2 teaspoons cumin seeds and 1 teaspoon coriander seeds using a pestle and mortar. Lightly toast in a dry frying pan for 30 seconds. Tip into a bowl and add a deseeded and finely chopped, medium strength red chilli, 2 finely chopped spring onions, 1 crushed garlic clove, 2 tablespoons finely chopped parsley, fennel or coriander and 4 oz (125 g) softened lightly salted butter. Beat well to mix and turn into a small serving bowl. Serve with smoked poultry, white fish, pork, or steak.

Fresh herb butter: Finely chop a handful of parsley, tarragon, chives, and the leaves from several thyme sprigs. Place in a bowl with 2 crushed garlic cloves, plenty of freshly ground black pepper, and 4 oz (125 g) softened, lightly salted butter. Beat well to mix and turn into a small serving bowl. Serve with any fish, poultry, or game.

Hot-smoked spiced turkey

This can be your smoking signature dish—a whole boned turkey, stuffed with tomato and coriander-flavored sausage meat, shaped into a neat "pillow" and tied with string to secure. Order from the butcher and ask him to bone the turkey but leave whole. It's a joy to serve as there are no bones to maneuver around; instead, the meat is carved into chunky slices that contain white and brown meat, plus the delicious stuffing. Serve with an herb salad and a fresh tomato sauce.

Serves 8

11–13 lb (5–6 kg) turkey, boned

Salt and freshly ground black pepper

FOR THE STUFFING

2 tbsp olive oil

1 large onion, finely chopped

3 garlic cloves, finely chopped

1 tsp dried crushed chillies

1 lb (450 g) good quality sausage meat

3 oz (85 g) sundried tomatoes in oil, drained and chopped

1 oz (25 g) coriander, chopped

3½ oz (100 g) pine nuts, lightly toasted

1 To make the stuffing, heat the olive oil in a frying pan and gently fry the onion for 5 minutes until softened, adding the garlic and chilli for the last couple of minutes. Leave to cool.

2 Put the sausage meat into a bowl and add the onion mixture, tomatoes, coriander, pine nuts, and a little seasoning. Mix thoroughly together. This is easiest done with your hands.

3 Lay the turkey out on the work surface, skin face down, and cut away any fatty areas around the skin and coarse sinews around the leg meat. Season all over with salt and pepper.

4 Tip the stuffing out onto the
meat and pack it down
neatly. Bring the edges of
the turkey up and over
the stuffing to enclose it
completely. You might
find that the meat
keeps a better shape if
you fold ends of the turkey
over the filling and then the sides to
make a parcel shape. Tie the turkey at 1¾ in (4 cm) intervals with string.
If the ends feel like they might open, then secure a couple of pieces of
string lengthwise. Turn the turkey over so all the joins are underneath.
Don't worry if it is a slightly irregular shape, it will still look delicious when
cooked. Weigh the meat to calculate cooking time, allowing 1 hour for
every 1 lb 2 oz (500 g).

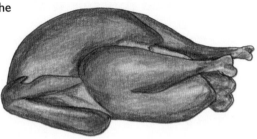

5 Spray a large square of muslin with water and use to wrap the turkey
up. Alternatively, put the turkey in a length of muslin stockinette and tie at
the ends with string. Spray the muslin with water.

6 Hot smoke the turkey for the calculated time. Unwrap about 1 hour
before the time is up to check the color. It should be a deep golden
brown. Return to the smoker unwrapped (or re-wrap if you don't want it
to color any more) and cook for the remaining time. Test by piercing the
meat with a skewer—the juices should run clear. Leave to stand for 30
minutes in a warm place before removing the string and slicing

Hot-smoked mussels with romesco sauce

Mussels absorb the flavor of smoke very easily. Fresh clams, as long as they not too small and fiddly, also work very well.

Serves 6

2 lb 6 oz (I kg) fresh mussels

2 tbsp olive oil

2 shallots, finely chopped

I tbsp chopped thyme

Squeeze of lemon or lime juice

FOR THE SAUCE

3½ oz (100 g) blanched almonds, roughly chopped

6 tbsp olive oil

2 oz (50 g) white bread, crumbled

2 red peppers, deseeded and roughly chopped

I medium-strength red chilli, deseeded and chopped

2 garlic cloves, crushed

I–2 tbsp sherry vinegar or white wine vinegar

Salt

Chopped coriander, to sprinkle

I First make the sauce. Lightly toast the almonds in a dry frying pan. Tip into a food processor and blend until finely ground.

2 Heat half the oil in the frying pan and lightly fry the bread until golden. Drain to the food processor. Add the red peppers to the pan and fry very gently, stirring frequently until very soft. Add the chilli and garlic and cook for a further minute.

3 Tip the pepper mixture into the food processor and add I tbsp of the vinegar and the remaining oil. Blend to a thick paste, scraping the mixture down from the sides of the bowl. Check the flavor, adding a little more vinegar for extra tang and seasoning with a little salt if necessary. Turn into a bowl and chill until ready to serve.

4 To prepare the mussels, clean them in cold water, discarding any damaged shells or any open ones that do not close when tapped firmly against the side of the sink.

5 Heat the oil in a large saucepan and gently fry the shallots to soften, about 5 minutes. Add the thyme and a squeeze of lemon or lime juice. Tip in the mussels and cover with a lid. Cook for about 5 minutes, shaking the pan frequently until the mussels have opened. Stir with a wooden spoon to combine the mussels with the shallots and ladle out onto a wire rack that fits inside the smoker. Discard any mussels that have not opened.

6 Spread out on the rack and hot smoke for about 30 minutes or until the mussels have absorbed the smoky flavor. (Check one first before removing them all from the smoker). Tip into a serving bowl, scatter with the coriander, and serve with the sauce.

Hot-smoked venison with bramble sauce

Venison is a smoking favorite. Its rich, gamey flavor is lovely with sweet, fruity sauces.

Serves 4

4 loin of venison steaks

Salt and freshly ground black pepper

9 oz (250 g) wild berries

5 tbsp sloe gin or cassis

2 oz (50 g) lightly salted butter

2 shallots, finely chopped

Pinch of ground cloves

1 Season the venison steaks on both sides with salt and pepper. Place on the smoker rack and hot smoke until cooked to your preferred taste. If you like it still pink in the center, test after about 25 minutes. For venison that is cooked through, smoke for about 35–40 minutes.

2 While smoking the meat, make the sauce. Reserve a handful of the berries and blend the remainder in a food processor with the sloe gin or cassis to make a purée. Press through a sieve into a bowl.

3 Melt half the butter in a frying pan and gently fry the shallots for 3–4 minutes until softened. Add the cloves and fruit purée, and cook very gently for 5 minutes until very hot. Remove from the heat and whisk in the remaining butter. Stir in the reserved berries and serve with the venison.

Moroccan spiced hot-smoked chicken

Pushing a flavored butter under the skin of chicken keeps it moist and succulent during smoking. Serve with steamed couscous or buttered new potatoes and an herby salad.

Serves 4

2 preserved lemons

1 oz (25 g) butter

1 onion, finely chopped

3 garlic cloves, crushed

2 tsp ras el hanout spice

Handful of fresh coriander, chopped

Salt and freshly ground black pepper

3 lb 5 oz (1.5 kg) chicken

1 Halve the preserved lemons. Scoop out and discard the pulp and finely chop the skin. Melt the butter in a frying pan and gently fry the onion for 5 minutes until softened but not browned. Add the garlic and spice and fry for a further minute. Tip into a bowl and stir in the coriander, lemon, and a little seasoning. Leave to cool.

2 Rinse out the chicken and pat dry with paper towels. Lift up the skin over the breast meat and slide your fingers carefully between the skin and meat to create space for the stuffing, taking care not to break the skin. Work as far as you can under the skin, if possible releasing it from the tops of the legs as well.

3 Spoon the spice mixture under the skin and spread out in an even layer making sure a little goes over the leg meat. Tie the legs together with string.

4 Place the chicken on a smoker rack or large wire rack that fits inside the smoker and hot smoke for 3 hours, or until the juices run clear when the thickest area of the thigh is pierced with a skewer. Leave to stand in a warm place for 20 minutes before carving.

Hot-smoked pork and chive sausages

Like all pork recipes, sausages are quick to absorb a smoky flavor. If you are making these sausages for anyone who cannot eat wheat, substitute 2 oz (50 g) flaked rice steeped in boiling water for 10 minutes then drained, instead of the breadcrumbs.

Makes 2 lb 12 oz (1.2 kg)

15 oz (950 g) good quality pork sausage meat

2 oz (50 g) pork fat, minced

3½ oz (100 g) breadcrumbs

2 oz (50 g) snipped chives

2 tsp salt

Freshly ground black pepper

Sausage casing

1 Put the sausage meat, pork fat, breadcrumbs, chives, salt, and plenty of freshly ground black pepper in a large bowl. Mix together thoroughly so all the flavors are evenly combined. This is easiest done with your hands.

2 Take a small ball of the mixture and flatten into a cake. Fry or microwave and taste for seasoning, adding a little more salt if liked.

3 Soak the sausage casing in water for about 15 minutes until softened. Drain and thread one end onto the 1 in (2.5 cm) wide nozzle of a sausage machine, leaving the end of the casing untied. Load the sausage machine with the meat mixture and slowly squeeze it though with one hand to fill the casing, supporting the casing as it fills with the other hand so it doesn't buckle up. Once filled to the required length, remove from the nozzle, leaving plenty of the casing ends left for tying. Smooth the casing out to check that they are evenly filled and that there are no air bubbles. If there are air bubbles, use your fingers to smooth them along the casing. Twist the sausages at 4 in (10 cm) intervals, or the length you prefer. Tie the open ends into knots, as close to the filling as you can. Cut off the excess casing with scissors.

4 Place on a smoker rack and hot smoke for about 50 minutes or until cooked through. Test by halving one sausage before removing them all from the smoker.

Variation

For pork, onion, and thyme sausages, heat 1 tbsp vegetable oil in a large frying pan and gently fry 2 finely chopped onions with 1 tsp caster sugar until soft and deep golden. Stir in 2 tbsp freshly chopped thyme, turn into a bowl, and leave to cool. Make the sausages as above using the onion and thyme mixture instead of the chives.

Salt and pepper smoked squid with garlic and herb mayonnaise

Squid is great hot smoked and perfect for summer eating. Ideally you need squid tubes that are about 2¾ in (7 cm) long so they don't fall through the rack.

Serves 4

1 lb 2 oz (500 g) small squid, cleaned

½ tsp freshly ground black pepper

½ tsp salt

1 tbsp olive oil

FOR THE MAYONNAISE

2 egg yolks

½ tsp Dijon mustard

1 garlic clove, crushed

1 cup oil (use either sunflower oil or a mixture of mild olive oil and sunflower oil)

1–2 tbsp white wine vinegar

Small handful roughly chopped herbs, e.g., parsley, coriander, fennel, lemon thyme

Salad leaves, to serve

1 Rinse and dry the squid and pat dry on paper towels. Put the pepper, salt, and olive oil in a bowl, add the squid and mix well. (If the heads and tentacles were included with the squid, cut the tentacles from the heads, discard the heads, and add the tentacles to the bowl).

2 To make the mayonnaise, put the egg yolks, mustard, garlic, and a little seasoning in a food processor or blender, and blend briefly to mix. With the machine running, gradually add the oil in a thin trickle until thickened.

3 Add 1 tablespoon of the vinegar and blend briefly again. Add the herbs and blend briefly to mix.

4 Season to taste adding a little more vinegar for extra tang, if liked. Turn into a small serving dish and chill until needed.

5 Place the squid on a fine mesh rack, or line a rack with foil if you think they might fall through, and space the squid on top. Hot smoke for about 45 minutes or until cooked through. Serve with salad leaves and the mayonnaise.

Variation

For an alternative dipping sauce for the squid, try this fiery chilli one. Finely chop 1 red chilli and a ¾ in (2 cm) piece of fresh ginger. Mix together in a bowl with 2 tbsp caster sugar, 1 crushed garlic clove, juice of 1 lime, 1 tbsp fish sauce, and 1 tbsp light soy sauce. Turn into a small serving dish and chill until needed.

Spicy pork spare ribs

This recipe shows how you can use smoking to enhance conventionally cooked meat. Tenderizing pork ribs in the oven, then finishing them off in the smoker gives then such an authentic flavor, but still juicy texture.

1 Put the ribs in a shallow, non-metallic dish and spread out in a single layer. Beat together the garlic, maple syrup or honey, vinegar, tomato paste, orange rind, and paprika.

2 Pour over the ribs and cover loosely with foil. Chill for several hours or overnight.

Serves 4

2 lb 12 oz (1.25 kg) meaty pork spare ribs

2 garlic cloves, crushed

3 fl oz (75 ml) maple syrup or runny honey

3 tbsp red or white wine vinegar

3 tbsp tomato paste

Finely grated rind of 1 orange

2 tsp hot paprika

Salt

3 Preheat the oven to 350°F/180°C. Turn the ribs into a roasting tin and pour over the marinade juices. Sprinkle with a little salt and bake for 1¼ hours, basting frequently with the juices.

4 Once the juices have thickened and are beginning to turn quite sticky, remove the ribs from the oven and transfer to the smoker rack, reserving the juices. (Alternatively place on a wire cooking rack that fits inside the smoker.)

5 Hot smoke the ribs for a further hour, basting frequently with the reserved juices or until the ribs are deep golden.

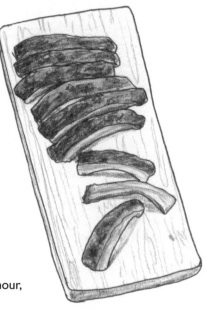

Variation

For spare ribs with a slightly Asian flavor, try this marinade and use as above. Mix together 2 crushed garlic cloves, 2½ fl oz (75 ml) honey, 1 oz (25 g) finely chopped fresh ginger, 3 tbsp rice wine vinegar, finely grated zest of 1 lime, 2 tbsp soy sauce, 1 tbsp allspice and 2 tbsp tomato paste.

Eggs benedict

This delicious breakfast/brunch dish is a good way to show off your perfectly formed ham.

1 Make the sauce first. Cut the butter into pieces and put in a small saucepan. Heat gently until completely melted, skimming off any white foam that settles on the surface. Pour into a bowl, leaving the white sediment in the base of the pan.

2 Put the vinegar, bay leaf, peppercorns, and 1 tablespoon water in a small saucepan and heat until bubbling. Cook until the liquid has reduced to about 1 tablespoon, watching closely as the pan will quickly run dry.

Serves 4

4 eggs

4 English muffins

8 slices home cured ham

FOR THE SAUCE

7 oz (200 g) lightly salted butter

2 tbsp white wine vinegar

1 bay leaf

½ tsp black or white peppercorns

3 egg yolks

Squeeze of lemon juice

Salt and freshly ground black pepper

3 Strain into a heatproof bowl and add the egg yolks, whisking with a balloon whisk to mix. Set the bowl over a saucepan of gently simmering water, making sure the base of the bowl is not touching the water.

4 Turn off the heat but leave the bowl over the hot water and very slowly pour in the warm melted butter, whisking constantly as you pour. The mixture will gradually thicken to make a smooth buttery sauce. Add a squeeze of lemon juice for a little extra tang and season to taste with salt and pepper.

5 Poach the eggs in boiling water. Split and toast the English muffins and place on warmed serving plates. Top with the ham slices and poached eggs and pour over plenty of sauce to serve.

Pot roasted bacon with lentils

Bacon and lentils make perfect partners. Don't forget to taste the bacon once cured by cooking a thin slice to check for saltiness. You might need to soak the joint in cold water for several hours or overnight first.

Serves 6

8 oz (225 g) Puy lentils

2 tbsp olive oil

2 onions, thinly sliced

4 garlic cloves, finely chopped

10 oz (300 g) cherry tomatoes, halved

3 lb 5 oz (1.5 kg) joint of brine cured bacon e.g collar, shoulder or loin

Several sprigs of rosemary and thyme

7 fl oz (200 ml) dry white wine

2 tbsp capers, drained and rinsed

6 anchovy fillets, roughly chopped

Freshly ground black pepper

Creamy mashed potatoes, to serve

1 Preheat the oven to 350°F/180°C. Rinse the lentils in a sieve and tip into a saucepan. Cover with water and bring to the boil. Simmer gently for 15 minutes, then drain and reserve.

2 Heat the oil in a large, flameproof casserole and gently fry the onions for 5 minutes to soften. Add the garlic and tomatoes and cook for 2–3 minutes to soften. Spoon the tomatoes onto a plate and reserve.

3 Add the bacon to the pan and cover with a lid. Bake in the oven for 1¼ hours.

4 Scatter the lentils and herbs around the meat and add the wine. Stir in the capers and anchovy fillets, and season with pepper. Return to the oven for a further 45 minutes, adding the tomatoes after 20 minutes. Leave to stand for 15 minutes before carving.

Mediterranean salt fish

Home-salted fish has a flavor that reminds us of summer eating, somewhere hot. This one-pot supper is even good in the winter.

1 Cut the salt fish into chunks, discarding the skin and checking for any stray bones.

2 Heat the oil in a large saucepan and gently fry the onion and fennel until softened, about 5 minutes. Add the garlic and fennel seeds and cook for a further minute.

3 Add the stock and saffron, crumbling the threads into smaller pieces. Heat until the stock is gently simmering then stir in the potatoes and salt fish. Cook very gently for 15–20 minutes until the fish and potatoes are tender.

Serves 4

1 lb 2 oz (500 g) salt fish, soaked in water for 24–48 hours

4 tbsp olive oil

1 small onion, chopped

1 small fennel bulb, chopped

3 garlic cloves, crushed

1 tsp fennel seeds, lightly crushed

2 cups (600 ml) fish stock

1 tsp saffron threads

1 lb 10 oz (750 g) waxy potatoes, cut into chunks

14 oz (400 g) small tomatoes, skinned and quartered

5 tbsp roughly chopped parsley

Salt and freshly ground black pepper

Warmed baguette or ciabatta, to serve

4 Stir in the tomatoes and parsley and cook for a further 5 minutes. Season with plenty of black pepper and check the seasoning. You might not need any salt. Ladle into bowls and serve with warmed bread.

Smoked chicken panzanella

Panzanella is a delicious Italian salad in which torn bread soaks up the garlicky tomato dressing. Adding smoked chicken turns it into a main meal version.

Serves 4

1 lb 2 oz (700 g) small, well-flavored tomatoes, skinned

6 oz (175 g) piece ciabatta

14 oz (400 g) smoked chicken breast fillets

Handful of pitted black olives

1 small red onion, finely sliced

Large handful of basil leaves

4 fl oz (125 ml) extra-virgin olive oil

2 garlic cloves, crushed

Approx 2 tbsp red wine vinegar

Salt and freshly ground black pepper

1 Quarter the tomatoes. Using a teaspoon scoop, out the seeds and pulp into a sieve set over a bowl. Press the pulp with the back of a spoon to squeeze out the juice. If the pulp is quite firm you can whiz it in a food processor first so it's easier to extract the juice.

2 Tear the bread into small pieces and scatter in a salad bowl. Thinly slice the chicken and add to the dish along with the olives, onion, and tomato flesh. Tear any large basil leaves into smaller pieces, discarding the stalks, and add all the leaves to the bowl.

3 Whisk the strained tomato juice with the oil, garlic, and 1½ tbsp of the vinegar so the dressing is tangy but not too sharp. Add a dash more vinegar if necessary. Season with salt and plenty of freshly ground black pepper. Pour the dressing over the salad and leave to stand for about 15 minutes so the bread partially absorbs the garlicky dressing.

Smoked mackerel pâté

Simple and delicious, whizzing home-smoked mackerel into a pâté makes a change to serving it freshly hot smoked.

1 Skin the fish, tear into flakes, and put in a food processor. Lightly crush the peppercorns using a pestle and mortar and chop the gherkins. Add the peppercorns and gherkins to the food processor and blend lightly to mix.

2 Melt the butter and pour half into the processor. Add the mascarpone and blend until mixed, scraping the mixture down from the sides of the bowl with a spatula. Add the lemon juice and season to taste with plenty of black pepper and a little salt if needed.

3 Pack into a serving dish, or several individual ones, and press down with the back of a spoon. Drizzle with the remaining butter and chill until ready to serve. Serve with warm toast.

Serves 4–6

14 oz (400 g) hot smoked mackerel

2 tsp green peppercorns in brine, rinsed and drained

4 small gherkins

2 oz (50 g) lightly salted butter

7 oz (200 g) mascarpone cheese

Squeeze of lemon juice

Salt and freshly ground black pepper

Toast, to serve

Suppliers

Meat Processing Products
www.meatprocessingproducts.com
PO Box 5755
Incline Village, NV 89450
Meat Processing Products sells meat preparation and butcher supplies, dehydrators and accessories, food slicers, marinades, pumps, mixers, injectors, sealers, saws, and much more.

The Sausage Maker Inc
www.sausagemaker.com
1500 Clinton St, Bldg 123
Buffalo, NY 14206-3099
The Sausage Maker sells butchering tools, casings and accessories, dry-curing products, food scales, kitchen appliances, meat grinders and storage equipment, tenderizers, slicers, smokehouses, and more.

Leifheit
www.leifheitus.com
510 Broadhollow Road, # 201
Melville, NY 11747
Leifheit sells carafes, jars for preserving and storing, kitchen scales and other appliances, and other household necessities.

Smoke'n'Fire
www.smokenfire.com
8030 W. 151st Street Overland Park
KS 66223
Smoke'n'Fire specializes in grills and smokers, charcoal and pellets, fire starters, smoking wood, barbecue accessories, and other barbecue-related novelties.

Smokinlicious
www.smokinlicious.com
110 North 2nd Street
Olean, New York 14760
Smokinlicious provides wood chips and wood blocks for smoking, fire starters, and smoking starter kits.

Cabela's
www.cabelas.com
5225 Prairie Stone Pkwy.
Hoffman Estates, IL 60192
In addition to an extensive list of outdoor goods, Cabela's sells grills, smokers, BBQ equipment and accessories, cookers, and marinade injectors.

Walton Inc.
www.waltonsinc.com
430 N. Mosley St
Wichita, KS 67202
Walton Inc. offers a wide selection of grinders, smokehouses, scales, sausage stuffers, mixers, tumblers, and grilling accessories.

Index